# BUILDING SAFE SYSTEMS IN AVIATION

# Building Safe Systems in Aviation
## A CRM Developer's Handbook

NORMAN MACLEOD
*Director of Turboteams, UK*

ASHGATE

Published by
Ashgate Publishing Limited
Gower House
Croft Road
Aldershot
Hampshire GU11 3HR
England

Ashgate Publishing Company
Suite 420
101 Cherry Street
Burlington, VT 05401-4405
USA

Ashgate website: http://www.ashgate.com

**British Library Cataloguing in Publication Data**
MacLeod, Norman
   Building safe systems in aviation : a CRM developer's
   handbook
   1.Flight crews - Training of 2.Aviation ground crews -
   Training of 3.Aeronautics - Human factors 4.Aeronautics -
   Safety measures
   I.Title
   387.7'3

**Library of Congress Cataloging-in-Publication Data**
MacLeod, Norman, 1954-
   Building safe systems in aviation : a CRM developer's handbook / by Norman MacLeod.
         p. cm.
   Includes index.
   ISBN 0-7546-4021-3
   1.  Flight crews. 2.  Human behavior. 3.  Personnel management. 4.  Aeronautics--
Safety measures. 5.  Aeronautics, Commercial--Employees. I. Title.

   TL553.6.M33 2005
   629.13'092--dc22

                                                                                    2005011794

ISBN-10: 0 7546 4021 3

Printed and bound in Great Britain by MPG Books Ltd, Bodmin, Cornwall

# Contents

*List of Figures* — vii
*List of Tables* — ix
*Preface* — xi

## PART I: ESTABLISHING THE AIM

1   Defining Crew Resource Management — 3

2   Safety and the Learning Organisation — 11

3   Establishing the Goal – Identifying CRM Behaviour — 31

## PART II: THE CONDUCT OF TRAINING

4   Developing Training Activities — 61

5   Delivering Training — 109

## PART III: MEASURING RESULTS

6   Measuring the Effectiveness of CRM Training — 125

7   Measuring CRM Skills — 143

8   Administration of the Process — 161

*Index* — *171*

# List of Figures

2.1 Boeing 727 Water Flask                                              12
2.2 The Contradictions in Safety                                        21
2.3 Creating Knowledge                                                  26
3.1 Behavioural Control                                                 40
3.2 How 'Culture' Gets Made                                            44
4.1 Topic Concept Map                                                   68
4.2 Desert Crash Opening Moves                                          82
6.1 Basic Evaluation Strategy                                          132
6.2 Training Transfer Evaluation Strategy                              132
6.3 Evaluation and Return on Investment Strategy                       133
6.4 The Benefit Stream                                                 137
7.1 Sampling Performance                                               146

# List of Tables

2.1 Barrier Analysis 14
2.2 Go-arounds in a UK Airline 24
3.1 Expertise Matrix 38
3.2 Training Decisions Matrix 55
4.1 Facilitator Competences 61
4.2 The 8 Step Process for Course Production 62
4.3 Workload Management 67
4.4 Strengths and Weaknesses of Classroom Activities 69
4.5 Training Inputs and Outputs 70
4.6 Media Selection 71
4.7 Media Production Times 72
5.1 The Introduction 110
5.2 The Events of Instruction 111
5.3 The UK CAA CRM Instructor Competencies 118
6.1 Training Cost Calculator 134
6.2 Training Programme Cost Calculator 135
6.3 Short-term Performance Measures 138
6.4 Long-term Performance Measures 139
7.1 CRM Skills Assessed in Airline 151
7.2 Grade Scales Used in Airline 151
7.3 NOTECHS Grade Scale 153
8.1 Change Management Activities 161

# Preface

An essential skill required of instructors in most areas of industry and commerce is the ability to swim. I say this because all too often I see able and enthusiastic individuals thrown in at the deep end with little support and no direction as to where safety might lie. This book is intended as a temporary buoyancy aid for those facilitators given the task of delivering Crew Resource Management (CRM) training.

In this very first paragraph I have managed to use the words 'instructor' and 'facilitator', alas not in the same sentence. The debate about CRM, generally, has generated more than its fair share of hot air and, specifically, the confusion over what to call the person at the front of the class is a good example of the muddled thinking detectable in the field. Re-badging does not improve the quality of the product delivered. The role of the instructor/facilitator is to promote learning. What you call the person matters little; what they do on behalf of their class is what counts. In this book I try to get people to think about the reason why we call groups of people together to discuss safe operations, what form training could take, how to turn ideas into action and how to measure the results.

Needless to say, what follows represents the distillation of many years experience of trying to deliver CRM training across a variety of operations and in many different countries. I have always taken encouragement from the fact that, despite my efforts, the people I have worked with have usually been interested in the subject, see its value and recognise the need to improve the way in which aviation 'works' if safety is to be guaranteed. Some readers may recognise their input in the pages that follow so it is only fair to offer my thanks to all those who have allowed me to experiment on them in the name of progress. However, some individuals have also had to endure close questioning and hypothetical debate about what CRM actually means and to Jerry Bresee, Katherine Senko, Giles Hammond, and Martin Pletscher I can only offer my apologies for some dumb questions. I must also thank the staff at Ashgate for their support throughout the difficult birth of this book and John Hindley, in particular, for never once collapsing with laughter as yet another forecast for delivery of the manuscript was proposed.

The views expressed in this book are my own and any apparent criticism was probably intended; just not disguised well-enough.

Norman MacLeod
Buckden, Cambridgeshire, March 2005.

# Part I
# Establishing the Aim

# Chapter 1

# Defining Crew Resource Management

## Birth of an Industry

On 28 December 1978 a United Airlines DC-8, Flight 173, crashed into woodland near Portland, Oregon. The subsequent NTSB accident report cited poor communication within the flight crew as a causal factor. And so a new industry was born: Crew Resource Management (CRM) training. Of course, the idea did not burst fully-fledged onto the scene. An earlier accident in which an aircraft crashed on take-off because the crew had failed to select flaps had prompted discussions between the FAA and the US aviation industry. This, in turn, led to the framing of new training requirements.

United Airlines, in 1981, were the first to run a training course that could clearly be identified as falling under the umbrella of CRM. Other airlines had, for years, been touching on aspects of CRM albeit under the heading of airmanship or captaincy. Pan Am introduced 'crew concept training' in 1974 but, having mandated the crews to operate as effective teams, the airline failed to offer any guidance as to how this goal might be achieved (Helmreich and Foushee, 1993). Now, for the first time, the crew as a social unit was the subject of discussion. Since then, CRM has developed beyond recognition. The subject is now a mandatory requirement in many countries. The scope has been extended beyond the flight deck to include cabin crew, dispatchers and maintenance personnel. Requirements for minimum standards of competence for those involved in the delivery of training are beginning to emerge. Even the name has changed. What started off as crew co-operation or co-ordination has morphed through cockpit resource management, via 'crew' to the current flavour, company resource management.

At the same time, developments in aircraft technology over the period have thrown up associated problems. The rapid growth of automation has resulted in questions being asked about the effectiveness of the methods applied to crew training in this area. The recognition that the relationship between the human and the technology installed in aircraft is a legitimate area of study, and that crew failures in one area can cascade throughout the work process, has resulted in the term 'human factors' being added to the training vocabulary. We now see human factors being treated, on the one hand, as synonymous with CRM and, on the other hand, as some sort of precursor to CRM. Given this confusion, it seems sensible to clearly establish the boundaries of CRM, at least as they appear to me. However, before I do that, it may be of benefit to examine views already proposed and which have achieved a general level of acceptance.

## What's in a Name?

Let us start with CRM itself. John Lauber, a psychologist then serving as a member of the NTSB, defined CRM as 'using all available resources [...] to achieve safe and efficient flight operations' and this is the definition trotted out in classes around the world to this day. Historical events have caused us to tinker with the initial letter of the acronym as I mentioned earlier. So, the accidents at Dryden, Canada and Kegworth, England (both in 1989), in which cabin crew were later found to possess valuable information that may have resulted in a more benign outcome had it been passed to the flight deck, shifted the emphasis from cockpit to crew. More recently, the acceptance of the fact that accidents are usually very fuzzy problems involving actors off-board as well as on-board the aircraft has shifted the focus to the system, represented by the company. Of course, any high school business studies student will tell you that a company is a socio-technical structure designed to manage resources to achieve a return on investment. As such, Company Resource Management, if not quite an oxymoron, is probably a redundancy that throws little light on the subject. What we know from accident and incident reports is that crews consistently fail to 'use all available resources'. By introducing the company, however, we are recognising that aviation is an economic activity. By broadening the scope of the definition we are also accepting that aviation is a system embedded in a complex web of relationships. So, the standard definition really doesn't help much in that it says little about what resources we are trying to manage, why we fail to make use of those resources, what management skills are required of aviation personnel and who, specifically, is supposed to be doing the managing.

Helmreich, Merritt and Wilhelm (1999) reviewed the history of CRM and have detected various 'generations' over which the subject has evolved until we reach the current iteration, which, in their view, is the age of error management. The standpoint they have adopted is clearly North American. For example, fourth generation CRM involved the proceduralization of CRM skills. This, in turn, is a requirement under the provisions of the FAA's Advanced Qualification Program (AQP). AQP only applies to the US and, even then, only on a voluntary basis. A significant number of airlines have dropped out of the AQP scheme rather than see the project through to conclusion. So, not all airlines, and certainly not all countries, are evolving in step. Equally, some of the first generation CRM techniques are still being widely, and successfully, used in training. The approach to CRM training has undoubtedly changed over time but it is hard to see how Helmreich *et als* 'generations' are anything other than an artefact of their analysis. The fact that CRM has gone through various iterations over time was first noted by Helmreich and Foushee (1993). However, to use an evolutionary model is probably a little too optimistic, implying, as it does, that each generation is somehow better adapted to its environment; better fitted to the purpose. Evolutionary history, of course, is littered with dead-ends: adaptations that did not work. After all, Neanderthals co-existed with *Homo sapiens* until the pressure of competition drove the Neanderthals to extinction. It is possible that we need a little more 'history' to accrue before we can distinguish the Neanderthal lines within CRM.

In an earlier attempt to clarify boundaries, Jensen (1997) places Aeronautical Decision making (ADM) at the centre of the problem. ADM seems to be an all-encompassing label, closer to 'expertise' in its meaning than decision making per se. Human Factors, according to Jensen, is the design of systems to take advantage of the characteristics and abilities of the operators while CRM is the application of ADM to multi-person flight crews. At this point we need to remember that Ohio State, where Jensen works, has done extensive research into pilot decision making in General Aviation, a fact reflected in the Jensen model. However, in Jensen's formulation, Human Factors is a domain in its own right. Initially, references to Human Factors were couched in terms that linked the topic to the relationship between humans and machines. However, Human Factors, too, has seen a rapid expansion over the lifetime of CRM. Wickens, Gordon and Liu (1997) define the goal of Human Factors as 'making the human interaction with systems one that: reduces error, increases productivity, enhances safety and enhances comfort'. Important here is the shift in thinking from the relationship between humans and devices to that of the relationship between individuals and systems. This line of thought is reflected in the Australian draft regulations (CASA, 2002) which define Human Factors as the 'relationship between people and their activities'. Given this definition, the subsequent definition of CRM as 'the application of Human Factors knowledge within the working environment' seems almost surplus to requirements. However, the CASA document picks up on current fashion and states that 'CRM deals directly with the avoidance of human errors and the management and mitigation of the consequences of those errors that do occur'.

Perhaps the real difference between Human Factors and CRM is that the former is a field of systems design whereas the latter is a response to systems failure. Human Factors attempts to optimise the design of the system, including the interface between the humans in the system and any technology they might use as part of their work. Human Factors principles can be applied after a failure in order to understand which elements of the design failed and why. CRM, however, has 'evolved' as each training initiative has either failed to deliver results or has been overtaken by events.

**Where are we Now?**

Returning to Helmreich's team at the University of Texas, in their review of CRM development they comment that broadening the scope of training to include personnel other than flight deck crew may have resulted in the 'unintended consequence of diluting the original focus on the reduction of human error'. Later on, when talking about fifth generation CRM they comment that 'Returning to the original concept of CRM as a way to avoid error, we concluded that the overarching justification for CRM should be error management'. It is true that concern was being voiced in the late 1970s that too many aircraft accidents were attributable to 'pilot error' but the placing of error centre-stage did not really start until James Reason's book, 'Human Error', was published in 1990. Interestingly, in the first major review of CRM published in 1993 (Wiener, Kanki and Helmreich), the work of Reason is cited just twice and even then only in the chapter on accident investigation (Kayten, 1993). Even in the final chapter, in which the editors review the future of CRM, reduction of human error is not mentioned.

In fact, in the early literature improved crew functioning is clearly the primary objective. Whether error management was the original goal of CRM training, as opposed to an outcome, and whether that message has been overlooked in recent years, is probably a moot point. However, to say that the overarching justification for CRM is error management is perhaps at risk of creating a false supposition. Various studies have put the incidence of major hull loss accidents that can be attributed to 'pilot error' at around 70-80%. Intuitively, then, if we manage errors better we should reduce the accident rate. The implications if this view will be discussed in more detail in the next chapter.

The regulatory authorities were comparatively slow to enter the fray. The FAA first issued guidance on the design and delivery of CRM in 1989 while the UK CAA followed suit in 1993. The FAA instructions differentiate between human factors, which is 'devoted to optimizing human performance and reducing human error' and CRM, which is the 'application of team management concepts' to 'address the challenge of optimizing the human/machine interface and the accompanying interpersonal activities'. Within the UK, CRM is 'a scheme that concentrates on crew members' attitudes and behaviour, and their impact on safety'. Within the European Joint Aviation Authority (JAA) framework we still see the emphasis being placed on communication and management skills rather than error management.

The IATA Human Factors Working Group (IATA 2001) published a paper that also sought to summarise developments in CRM by addressing the need for airlines to 'refocus Crew Resource Management towards the concept of error management as the 'tip of the arrow''. Confusingly, the paper describes CRM in its early days as a business management theory adapted to the flight deck. The bulk of the paper is a reworking of the Helmreich, Merritt and Wilhelm 'generations' interpretation. One item is of interest, though. Discussing the role of instructors, the paper argues that 'after decades of concentrating solely on error detection and correction, the shift to error management observers is a difficult one'. The paper also suggests that instructors will be required to give better grades to crews where an individual makes an error but, as a crew, the error is managed to the extent that there are no consequences. In fact, the paper confuses, at the very least, techniques of instruction with attitudes to failure (error) among instructors. However, importantly, the paper does accept that making errors is a by-product of performance in training and that dealing with errors is a legitimate, observable activity. In fact, it is the shift towards expecting CRM skills to be observed and measured that is the main thrust of regulatory involvement over the past few years.

## Broadening the Scope

So far, the discussion of CRM has revolved around the traditional areas of psychology and ergonomics, hardly surprising given that it is mainly psychologists who have been involved in CRM development. However, there are other disciplines that can throw light on problems in aviation. For example, Nichols (1997) criticise the focus on psychological models of accident and injury causation and suggests that progress cannot be made without understanding 'what determines the determinants of injury'. In this view, sociological models, which deal with human endeavour within systems of production, can offer insights into the motives of individuals. The rise of the low cost airline model has been lauded as a landmark in the development of the industry but few voices have been raised about the possible safety-related performance problems that could arise from the rigid application of the model. Anthropology, in the guise of comparative sociology, can give clues as to why human societies function the way they do. Airlines, after all, are societies and each airline differs slightly, one with another, in its internal workings. Because CRM is still relatively new, it is also relatively impoverished in its tools of interpretation and explanation. In order to fill some of the gaps I believe we will need to look outside of the traditional psychological/engineering models (see also Batteau 2002).

## Where Next?

Having given a very brief overview of the conceptual development of CRM, what interpretation underpins the rest of this book? In keeping with Wickens *et al*, we take a systematic view in which humans are free agents. We place the process of work firmly at the centre of the problem. Work is a collaborative process in which human actors apply technical and social skills in order to function effectively in achieving production goals. Safety and efficiency are outcomes of that process. However, the workplace is just one of many overlapping systems of which the human is a component. For example, most airline personnel participate in some sort of family life when they get the chance; all personnel are employees of an organisation; they are also citizens of their country. These overlapping systems will influence, directly or indirectly, the way work is done. And here is the key to our approach. The attempt to distinguish between CRM and Human Factors is really an attempt to identify distinct research domains. The shift towards error management could be interpreted as a frustrated response to the failure of training interventions to deliver an observable incremental improvement in safety. Both lines of argument fail to recognise that the real objective is to develop competence within individuals. By placing soft skills on an equal footing with the hard, stick-and-rudder, skills we are now able to identify a coherent competence framework that can be used to develop effective training systems.

Elsewhere (MacLeod, 2000) I have said that all training should start with an analysis of the task and of the training interventions best suited to achieve proficiency within the training resources available; perhaps not an original concept but one that is overlooked more often than not. This is the approach endorsed by AQP and the draft European counterpart, the Alternative Training and Qualification Programme (ATQP). By starting with a task analysis we are taking a bottom-up approach to CRM. But CRM is more than simply skills development. I see CRM as also serving the need of making sense of the workplace. CRM, for me, is both a lens and a toolkit. We view the workplace through the lens in order to understand why things happen. We use the toolkit in order to function more effectively in that workplace.

One of the criticisms of CRM has been its low transportability, a polite way of saying what works for you will probably not work for me. Therefore, organisations have to reinvent CRM in the light of their own operation, regulatory environment, employees and so on. This is, in effect, a recognition of the fact that, if not each individual airline, at least each sector of the industry has its own variation of the basic competence framework that will also need to be adapted to local conditions.

I want to state categorically that the primary goal of CRM training is to develop the social and cognitive skills that are exercised together with technical, systems-related, skills in order to achieve safe and efficient aviation. This is what I meant earlier when I said that the process of work is a central issue. But CRM is more than that, too. Because the process of developing competence frameworks requires analysis of an organisation, CRM can also act as top-down organisational development activity. I said earlier that safety is an outcome and in the next chapter I want to explore the idea of safety in more detail. However, safety is also clearly a goal. So, CRM is also a safety initiative. I have also commented on the need to view performance as production activity and, so, CRM is an efficiency tool. I will develop this production system idea in more detail in Chapter 3. If you accept my organisationally-based view, you will also have recognised that everything we discuss in this book applies equally to all work groups involved in the process of generating revenue through the operation of aircraft. The fact that regulations do not yet require baggage handlers, for example, to be exposed to CRM does not mean they are exempt from the problem. Ask yourself this: who does more damage, measured in cash terms, to aircraft in an average year, flight crew or ground handlers?

The remainder of the book will get to grips with the design and delivery of CRM training. In the rest of this section we will consider what we can learn from current formulations of safety as a concept. I then want to explore the process of work and how people learn about their jobs. Next, given that I have pinned my colours to the competence model of CRM, we will look at how we define the desired performance expected of crews. With our behavioural framework to hand, we will look at translating goals into activities designed to achieve those goals. We will examine in detail the methods available for delivering training before, in the final section of the book, I look at the problem of measurement, both in terms of effectiveness of training and in terms of behaviour on the line. My goal is to provide facilitators with a complete toolkit in order to support them in shaping CRM to meet their own company's needs.

## Conclusion

In this first chapter I have tried to trace the broad development of CRM and to suggest weaknesses in our current formulation. It is important to understand these weaknesses because our goal as training designers is do develop training interventions that show a return on the investment: our training needs to work in some way. So far, little of what has been done in the name of CRM training can be said to have delivered results. If we follow the narrow guidance on CRM training contained in published syllabi then there is every chance that what we offer will continue in the grand tradition of systematic impotence.

I have outlined the scope of the task that lies ahead. We are endeavouring to develop competent personnel who, through engagement with the work process, deliver safe and efficient workplace performance. Clearly, this goes beyond the narrow definition of CRM or Human Factors. The implication is that our efforts will impinge upon other company training initiatives – or even suggest training that currently does not exist. To fail to recognise that implication is to simply add to the reasons why your efforts will not deliver effective training. A harsh judgement, possibly, but that is the reality. Having wasted too much time worrying about what 'CRM' means I suggest you now expunge the term from your memory as being a brake on imagination and let us move on to something more interesting – Safety.

## What to do Next

In the light of the discussion so far take some time to reflect on your own view of CRM and its objectives. Ignore fads and buzzwords for a moment and try to establish a goal for a training intervention aimed at improving safety and operational efficiency. Ask yourself how is failure dealt with in your company? What are the implications for how training will be conducted in the light of corporate attitudes to failure? What measures of success exist in your company to tell management how well the workforce is performing?

## References

Batteau, A. (2002), 'Anthropological Approaches to Culture, Aviation and Flight Safety', *Human Factors and Aerospace Safety*, Vol. 2 (2), pp 147-172.

CASA (2002), *Human Factors and Crew Resource Management* casa.gov.au/nprm/nprm0211c09.pdf

Helmreich, R.L. and Foushee H.C. (1993), 'Why Crew Resource Management?', in E.L., Wiener, B.G. Kanki. and R.L Helmreich (eds), *Cockpit Resource Management*, Academic Press, San Diego.

Helmreich, R.L., Merritt, A.C. and Wilhelm J.A. (1999), 'The Evolution of Crew Resource Management Training in Aviation', *International Journal of Aviation Psychology*, Vol. 9 (1) pp 19-32.

IATA. (2001), *The Evolution of Crew Resource Management: From Managerial Theory to Safety Tool*, Human Factors Working Group.

Jensen, R.S. (1995), *Pilot Judgement and Crew Resource Management*, Avebury, Aldershot.

Kayten, P.J. (1993), 'The Accident Investigator's Perspective', in E.L. Wiener, B.G. Kanki and R.L Helmreich (eds) *Cockpit Resource Management*, Academic Press, San Diego.

MacLeod, N. (2000), *Training Design in Aviation*, Ashgate, Aldershot.

Nichols, T. (1997), *The Sociology of Industrial Injury*, Mansell, London.

Wickens C.D., Gordon, S.E and Liu, Y. (1997), *An Introduction to Human Factors Engineering*, Longman, New York.

Wiener, E.L., Kanki B.G. and Helmreich, R.L. (eds) (1993), *Cockpit Resource Management*, Academic Press, San Diego.

# Chapter 2

# Safety and the Learning Organisation

## Introduction

In the previous chapter I gave safety as a possible outcome from training, whether it be delivered under the banner of Human Factors or CRM. A concern with safety has been fundamental to aviation since its earliest days but because the technology we use to transport people has become ever more reliable, and therefore safer, the focus has shifted from the hardware to the people who control the hardware. Whereas in the last chapter I tried to show that our understanding of CRM was, perhaps, rather vague, in this chapter I want to see if our conceptualisation of safety is any more rigorous.

I want to explore more fully what we mean by safety, how safety is enacted in the workplace, how organisations learn about safety and, therefore, how safety can be promoted through training.

## What is Safety?

Simply put, safety is the freedom from, or lack of exposure to, unnecessary risk or harm (Turner, 1982). From this definition we can identify 2 possible indicators of an unsafe condition; first, a workforce that is being 'harmed' in some way and, second, a level of risk which exceeds some acceptable value. Measures of harm are perhaps easier to visualise. We can record workplace injuries and categorise them in terms of severity of injury or the time taken off work to recuperate. We can measure lost productivity or costs incurred in maintaining production. When we consider risk, however, we start to run into problems. Risk relates to an outcome or an event and has a number of components. First, we are interested in the probability of the event occurring. Second, we are interested in the possible consequences of the event in question. Finally, we are interested in the context of the event. I would like to illustrate these components in a real situation.

In this first example I want to examine the first two ingredients in our definition of risk: probability and consequence. A Boeing 727 reported an on-board explosion and made an emergency landing at Kai Tak Airport, Hong Kong. Investigators boarding the aircraft found the forward galley destroyed and the forward part of the cabin badly damaged. Fortunately, no passengers or crew had been harmed. The subsequent investigation traced the problem to one of the water heaters in the galley. The hot water flask, shown as a diagram in Figure 2.1, is fitted with a number of protective mechanisms. First, a thermostatic control regulates the temperature of the water in the flask. Second, the power cable to the heater is routed through a soldered connection on the outside of the flask. The solder will melt if the flask temperature exceeds a certain value, cutting off the power to the heater. Finally, a pressure relief valve is fitted to allow a build up of steam to vent to the cabin.

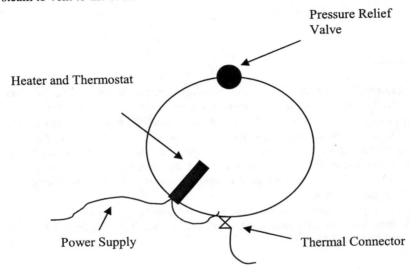

**Figure 2.1 Boeing 727 Water Flask**

The investigators discovered that each of the protective measures had been circumvented. First, the airline had shifted from scheduled maintenance for certain items, like thermostats in water flasks, to simply replacing items as they fail. Presumable, as part of this policy change, it had been assumed that a thermostat failure would render the flask inoperative as opposed to fully functioning, as was the case here. Furthermore, a failure would be detected in terms of a flask not working. On this day, the thermostat had failed but the crew did not yet know its condition. At some stage during previous maintenance the material used to solder the thermal connector had been replaced with a similar material but with a higher melting point. Thus, as the temperature rose in the flask – as a result of the faulty thermostat – the thermal connector remained intact, allowing power to continue to flow to the heating element. Finally, the flask had occasionally been filled with ordinary tap water rather than the deionised water stated in the instructions. As a result, over time, the pressure relief valve had become furred up, preventing the excessive pressure in the flask from being released. The result was, effectively, a bomb on the aircraft.

Each of the protective mechanisms in this device was breached by a different part of the organisation. The decision to change maintenance policy was taken by engineering management on the basis of an economic review.[1] The use of the wrong material for the thermal connector was, presumably, an error on the part of an individual technician. The use of tap water was an expedient on the part of the catering company or ground handling agent. Each act was removed in space and time from the explosive event. Each of the players in our short story would have had to consider the probability of an adverse event resulting from their action. However, consider, for a moment, the likelihood that anyone would have anticipated that a hot water flask could become an explosive device as a result of a change in maintenance procedures. Risk assessments require the use of imagination and, even then, actual observed outcomes are often unimaginable – that is what makes the whole question of 'risk' so problematical for mere mortals.

Of course, it is unlikely that any of our players even gave their actions a second thought but the story does throw light on another problem we face when we consider risk. Quite often, an adverse event is the result of several different lines of action coming together. Each of the breaches we have just seen would not, in isolation, have caused the outcome. Nor, in all probability, would any combination of 2 failures. But with all 3 defences broken – then we have a problem. However, each of the players in the scenario would only have considered their own individual act and would have been unaware of the part being played by others. So when we talk about risk, we have to recognise that it is almost impossible to consider all the possible permutations and interactions between elements in the system, even a system as simple as a hot water flask.

*Safety as Barriers*

I want to take a slight detour at this point as the water flask example provides a neat introduction to one model of accident investigation; that of barrier analysis (Hollnagel 1999). If you think of an accident as an uncontrolled release of energy with undesired consequences, then we can mitigate or prevent such an outcome through the use of barriers. Examples of barriers are given in Table 2.1. The water flask on an aircraft represents a physical barrier designed to contain hot water. The pressure release valve, similarly, is a physical barrier designed to dissipate surplus energy. The thermostat can be considered a functional barrier preventing the flask overheating, as is the thermal interlock in the power supply. The requirement to use deionised water would have been described in a procedure, a symbolic barrier, and the maintenance schedules represent immaterial barriers. Even though the designers of the system put various layers of defence in place to keep the user of the water flask safe, the barriers were penetrated on this occasion. Barrier analysis, in much the same way as the concept of defences (Reason 1990), forces us to explore the manner in which failure can propagate through a system and where in the system failure can, or should, be trapped.

My digression at this stage as intended to reinforce the earlier point about the complexity of inter-relationships, even in quite simple systems. When we apply a rigorous framework of analysis, such as barrier analysis, we can identify the ways in which we attempt to keep the system safe. At the same time, it can allow us to explore gaps, possible opportunities for breaches and generally poor design of barriers. However, objective frameworks tend to obscure the human involvement in the whole process and I want to return to our discussion of risk to explore how messy things get when humans are involved. First, though, to summarise, we said that risk involves probability and consequence. I have tried to show that it can be difficult, if not impossible, to anticipate the consequences of an action and, therefore, estimates of probability become meaningless. The problem is made worse by the coupled nature of technological systems (see Perrow, 1984) in which functional elements are inter-related in ways that are difficult to predict.

**Table 2.1 Barrier Analysis** (after Hollnagel)

| Barrier System | Barrier Function | Example |
|---|---|---|
| Material, Physical | Containing | Fuel tanks |
| | Restraining | Seat Belts |
| | Keeping together | Toughened glass |
| | Dissipating | Fire extinguishers |
| Functional | Preventing Movement | Locks, gates |
| Symbolic | Countering | Colour-coding of wires Labels and warnings. |
| | Regulating | Instructions, procedures. |
| | Indicating | Warnings, alarms. |
| | Permission | Flight plan, load sheet. |
| | Communication | Flight clearance |
| Immaterial | Monitoring | Checklists |
| | Prescribing | Orders and Instructions. |

*Back to Risk*

Returning now to risk, the final ingredient in our definition was the context in which the event occurs and we said that we could rephrase this as 'who says it's a problem?' In the late 1990s a number of court actions were brought against operators of the British Aerospace (BAe) 146 aircraft by cabin crew who claimed that their health had been impaired as a result of working on the aircraft. The matter came to a head in Australia after several pilots reported cases of in-flight incapacitation as a result of breathing contaminated air. Eventually, the Australian

Senate Rural and Regional Affairs and Transportation References Committee conducted an enquiry into the matter.

The BAe 146 was then in service around the world. In Canada, some cabin crew were pursuing their own claim. One Canadian airline had a small number of these aircraft that were being operated as a franchise for Air Canada. The airline had been equipped with turbo-prop aircraft and the step up to jets was bounded by pilot union-imposed constraints on aircraft numbers and the routes flown. These constraints, known as scope clauses, are not uncommon in situations where a large airline offers up certain routes to be flown by lower-cost regional companies. The cabin crew assigned to the BAe 146 noticed a decline in their health and many personnel were forced to leave their jobs as a result.

To fully understand the problem we need to take a technical detour. Cabin air is drawn from the compressor stages of the aircraft engines. It is filtered and cooled, mixed with air drawn from within the aircraft and then fed back into the cabin or the flight deck. The volume of air recycled over a given time period is greater on the flight deck than in the cabin. Air delivered to the cabin can contain contaminants drawn from outside the aircraft and also contaminants picked up within the aircraft. Engine oil, de-icing fluid and hydraulic fluids, for example, have found their way into the system. The proportion of air drawn from the engines can be varied. In the BAe 146, 2 settings are available 'Fresh' and 'Recirc'. The 'Fresh' setting does not mean that fresh air is being fed into the aircraft. It simply means that less cabin air is being mixed with the engine air than when 'Recirc' is selected. When in the 'Fresh' mode, more air is drawn off the compressor stages to feed the cabin pressurisation system and so the engines are working harder, and fuel consumption is increased accordingly. The condition of the filters installed in the system will also affect the purity of the air. The fact that air needs to be forced through filters, again, imposes a workload that, in turn, results in increased fuel burn. Anecdotally, I have been told that some airlines have remove filters in order to improve engine efficiency. So, airlines are able to make decisions about system maintenance, fuel burn, passenger comfort and staff health issues on the basis of a single factor; cost.

If we now return to the situation in Canada, the airline involved was aware of the possible health implications associated with operating the BAe 146. It had noticed the increase in sickness and staff turn-over. However, it was cheaper to pay sickness compensation claims to cabin crew than to change the way the aircraft was being operated. The pilots in this airline were aware of the problem and were sympathetic towards the plight of the cabin crew. However, under the conditions of the scope clause, the airline was not permitted to change the aircraft type. If the BAe 146s were disposed of, the pilots would have to return to flying the Dash-8 turbo-props. This would be a backward step in terms of both prestige and pay packets.

We can now analyse the BAe 146 case and see all 3 elements of the risk definition at work. We will start with the probability of an event occurring. In this case, we are interested in the chance that some failing in the design, maintenance or operation of the aircraft cabin conditioning system will have an unintended outcome. Of course, when the aircraft was designed it is probably fair to say that the estimated probability of the observed outcome was zero. Perhaps the problem began to emerge as aircraft aged and systems started to show signs of wear and tear. However, we have no data on the age of the aircraft involved in the incidents to support this view.

Our next criterion was the likely consequence should the event occur. In our example we can divide the consequences into 2 categories: the effects on the individual exposed to the risk and the effects on the aviation system. If we take the individual first, the possible consequences will be influenced by the degree of exposure, susceptibility to the effects of exposure, recovery periods between exposure and so on. Of course, it is entirely likely that individuals have been experiencing low-level symptoms of exposure for prolonged periods but have failed to recognise the problem. If we consider what happened in Australia, we see that there was a longer-term deleterious effect on over-all health of the cabin crew and a short term, almost-immediate incapacitation of some of the pilots. The latter time scale is easier to observe, of course, than the former and, given the critical stage of flight at which the pilots succumbed to the effects of breathing contaminated air, serves to magnify the apparent risk.

If we now turn to the system, we can identify a range of costs associated with the temporary replacement of sick workers, compensation for those permanently grounded, costs of maintenance or modification of the conditioning systems, increased fuel consumption and so on. I deliberately used the term system as opposed to airline as we need to consider the regulatory position. It may be that one consequence could be increase regulatory oversight or even demands for the conditioning system to be redesigned, at which point there are outcomes for the aircraft manufacturer.

The final criterion was the context within which the risk occurs or, to put it another way, the perspective from which the risk is viewed. Here we can identify many stakeholders. The company perspective would be one of risks set against cost of doing things differently. We can also analyse the situation in terms of the status of the various groups concerned. Cabin crew are still considered to be low-status employees who are readily replaced. Therefore, an airline need only employ cabin crew on short contracts for the problem to simply appear to disappear. We need to review the attitude of pilots in the Canadian airline who were aware of the problem and yet put personal self-interest above the health of their colleagues. That Canadian pilots are potentially at risk is borne out by the experience of their Australian BAe 146 colleagues.[2]

Part of the problem, as with the water flask, is that events do not occur in isolation and that an action can propagate across a network of inter-relationships. So, an airline choosing to remove filters or to defer maintenance can have an effect elsewhere in the system. Even the decision to allow smoking aboard an aircraft can affect the efficiency of the filtration system which, in turn, will affect the cabin air quality.

Before we can even begin to talk in terms of risk, we need to be aware (or recognise) that a problem actually exists. In the case of the BAe 146, we see that reports occurred in clusters. There is every indication that the problem had existed for some time but no action had been taken. It seems that a condition needed to exist above some threshold level in order for the risk to be recognised. It may be that the level of Trade Union activity in an airline or State may be the catalyst required for action to be taken. We need to ask, in the Australian situation, if action would have been taken if the health risk had been limited to cabin crew and not included pilots, as was eventually the case in this instance. The BAe 146 had been in service for many years before the question of crew health arose. Even after the Australian Senate Hearings the situation was summed up this way in a news report:

> The aircraft operators, Ansett, Qantas, National Jet, all say they've invested in solutions, but in the last three months, another 21 fume incidents have been reported, and, somehow, in between the conflicting technical and medical advice, these senators have got to make sense of something no-one appears to fully understand (ABC, 2000).

Risk, then, cannot exist in isolation; it needs a focus. Events or objects are risky but that property of 'riskiness' only emerges as a result of some interaction between actors. A surgeon's scalpel, for example, possesses the property of 'sharpness'. However, the sharp edge of the scalpel only becomes risk when in comes into close proximity to something or someone we don't want to cut. Of course, the actors in an event are not necessarily just the humans in the process. Any element coming into conjunction with other elements in such a way that it exerts influence constitutes an actor.

Having explored risk in some detail, I now want to look at how safety is enacted in the workplace. I started this chapter by saying that safety is a goal of training and, in the previous chapter I also alluded to safety being an artefact of the workplace. In the next section I want to explore the way in which people 'act safely'. We will kick of by looking at another fuzzy concept: safety culture.

## The Enactment of 'Safety'

Defining safety helps us to focus on the problem. However, our aim, as trainers, is to bring about change. Therefore, we need to understand how safety is enacted in the workplace: how do people behave 'safely'. We will explore the precursors of organisational behaviour in more detail in the next chapter but now is an appropriate time to look at the concept of safety culture. We know from statistical analysis that unsafe events are not evenly distributed across the aviation industry. If we compare airlines, accident rates can differ by a factor of 42 (Reason). If all airlines were equally exposed to the risk of an accident, then we would not expect to see such a large disparity. Quite often, an apparent higher than normal rate of unsafe events is ascribed to an airline's safety culture. When we move from looking at what safety is to how safety occurs, we usually start by considering how the organisation embodies safety.

We will see later that 'culture' is a problematic concept. Safety Culture is no different. In a review of attempts to define safety culture, Wiegmann et al (2004) traced the origin of the concept back to the Chernobyl nuclear accident in the Ukraine in 1986. However, we can go back further. In 1942 a university psychologist was enlisted into the Royal Air Force and was serving as an Aerodrome Controller – forerunners of today's Air Traffic Controllers (Paterson, 1955). Morale amongst aircrew on the base was bad and there seemed to be a link between morale and aircraft accident rates, in particular accidents on landing. In an attempt to reduce accidents a set of instructions was developed for approach and landing. These included statements such as 'You will commence the approach at 400 feet and fly a constant angle of descent', 'The speed of the approach is to be 95 mph for Spitfires', 'You are to touch down on the touch down mark'. A card index system was set up and the details of each pilot's landing technique were observed from the Control Tower and noted on the cards. At first, the system was dismissed by the pilots until an aircraft ran off the end of the runway. It was found that, according to the pilot's record card, he consistently landed long. By collecting data and feeding data back to the pilots, the 'culture of safety' in an operational wartime unit was changed, observed pilot errors halved and the accident rate was cut to a third. Paterson does not use the word 'culture' in his description of life on a wartime base. Instead, he talks about 'morale'. However, it is clear that he is addressing the same problems as safety managers today.

## Safety Culture

Although 'safety culture' was, possibly, given a name for the first time in 1986, we can be fairly sure that the matter of the relationship between how people view the world and how people act 'safely' in the world has been around at least since 1942. What, then, can we learn from how researchers have attempted to define safety culture since that first attempt after Chernobyl? In their paper, Wiegmann et al address two concepts, that of safety culture and of safety climate. Of safety culture they say:

- Safety culture is a concept defined at the group level or higher that refers to the shared values among all the group or organisation members.
- Safety culture is concerned with formal safety issues in an organisation and closely related to, but not restricted to, the management and supervisory systems.
- Safety culture emphasises the contribution from everyone at every level of an organisation.
- The safety culture of an organisation has an impact on its members' behaviour at work.
- Safety culture is usually reflected in the contingency between reward systems and safety performance.
- Safety culture is reflected in an organisations' willingness to develop and learn from errors, incidents and accidents.
- Safety culture is relatively enduring, stable and resistant to change.

Of safety climate they say:

- Safety climate is a psychological phenomenon that is usually defined as the perception of the state of safety at a particular time.
- Safety climate is closely concerned with intangible issues such as situational and environmental factors.
- Safety climate is a temporal phenomenon, a 'snapshot' of safety culture, relatively unstable and subject to change.

The authors attempt to situate culture and climate in much the same way as personality and attitude can be used to reference the extent to which an individuals' behaviour is consistent across time and place. Other authors have tried to characterise the enactment of safety in the sense of a disease in the system. In what is known as the Bio-medical model, disease is:

- An organic condition. Non-organic factors are excluded.
- A temporary organic state that can be eradicated by medical intervention.
- Something experienced by sick individuals who are then the object of treatment.
- Something that is treated after the symptoms appear. The application of medicine is a reactive healing process.
- Something that is treated in a medical environment away from the site where the symptoms first appeared.

If we apply the bio-medical analogy to aviation, we see that discussions of safety usually revolve around an individual who has performed in an unacceptable manner; this is the disease, the organic state of the victim. It is significant that in the Bio-medical model excludes non-organic factors; the problem resides in the patient. Of course, we are now beginning to accept that safety is the product of the interaction between individuals and the system of which they are a part, the key characteristic of our discussion of CRM in Chapter 1. In our analogy, the company safety officer represents the medical intervention and takes control as soon as 'deviation' is detected. The patient is taken off for interview. The collection and interpretation of data tends to be conducted away from the scene of the accident, usually by nominated safety officials ('doctors and nurses') and with the errant individual as the focus of attention (dealing with the symptoms). Quite often, the solution involves the individual being sent for 're-training' (the medicine).

Key to the bio-medical model as a conception of safety is as purely a management issue and, indeed, Wiegmann et al identify the fact that safety culture is seen as being related to the process of management and supervision. In the bio-medical model, an unsafe practice is a 'disease' in the system that needs to be treated. It is the responsibility of management to take action. Unsafe behaviour is considered to be a threat to production that can be solved, in part, by better job design or the provision of better preventative measures. In effect, safety is a planning task aimed at prevention and measured in terms of organisational loss.

Managers have a further responsibility to increase workers' commitment to safe working practices. In this model, the 'patient' is the object in need of treatment and yet the patient will also have a view, one that we will discuss later in this section. The review of safety culture, similarly, reflects a view that safety is somehow linked to a top-level view of what shapes workplace behaviour. Safety, then, is a management activity directed by the organisation to meet some goal defined by management – or is it. Perhaps we need to give the workers a voice.

## The View from the Inside

If we now move to consider the view of the 'patient' in our analogy we quickly come to the crucial question of who owns safety? If we consider the view from the workplace, what we find is that safety is a social construction (Rochlin). Safety, in this case, is a richer concept rooted in working practices. In this formulation, those that work in dangerous, changing environments learn to act in a safe manner, they do not learn safety (Gherardi & Nicolini). As we suggested in the first chapter, safety is an artefact of the manner in which working communities create ways of completing the task. Safety is an expression of trust among groups of people who share the same working environment. Trust is based on a perception of your colleagues' practical experience and their level of professionalism. In this case, professionalism means the degree of autonomy exercised by the individual and the level of responsibility expected of that individual. Safe working practices require an ability to predict variations in the work process and to understand their consequences. Safety involves monitoring your own work performance and that of others around you. Behavioural methods of safety management are based on the fact that the people who actually do the job know the most about it and, therefore, the problems to be overcome in getting the job done. The contradictions inherent in the discussion of safety are represented in Figure 2.2.

If we are to deliver effective training we need to resolve the contradictions reflected in the diagram. We will look more closely at the precursors of behaviour in the next chapter but for now it is enough to say that training attempts to shape observed behaviour. However, behaviour is exhibited by actors who, in turn, exercise agency; that is, they can choose the way they want to behave. As part of our training analysis, we need to take into consideration those factors that may prevent the execution of safe behaviour and, of course, company culture may be one of them.

An implication of this analysis is that what we have been calling safety culture is, in fact, something created in each workplace by each group of workers. Evidence of this was found in a study of railway track maintenance teams in Japan (Itoh, xxx). An analysis of measures of safety culture were correlated with accident rates and it was found that there were as many distinct 'safety cultures' as there were work groups in the study. The question this raises is that, if management do not control safety, then what role does the organisation play in promoting safety? Furthermore, how do we, as training designers, account for the organisational influence when developing courses aimed at changing workplace behaviour? We return to this in the next chapter.

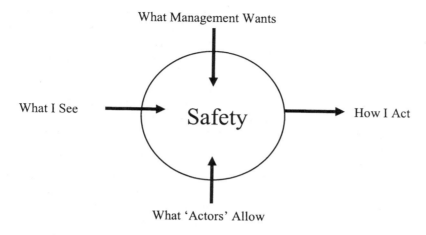

**Figure 2.2 The Contradictions in Safety**

**Safety as Risk Management?**

To conclude this section, and to pick up on the questions just posed, I want to offer a view of what the goal of CRM training might be. In the previous chapter I mentioned that 'error management' is, increasingly, seen as the ultimate – and original – objective of CRM. We now need to explore this concept in more detail and in the context of our examination of risk. The starting point for error managers is the extent to which the outside world contains 'threats'. A threat is any factor outside of the flight deck that can trigger an error on the part of flight crew. Not all threats give rise to errors and not all errors derive from a threat. An effective crew will, first, identify probable threats to their specific flight and develop ways of removing, controlling or mitigating their effects. Should these safeguards fail, then an effective crew will recognise any induced errors and will manage those errors to sustain, or return to, a safe flight condition. The analysis of crew effectiveness takes place within the procedural framework defined for the conduct of the task. So, current research largely involves cataloguing departures from procedures. By collecting data, the Threat and Error Management School can make judgements about, and comparisons between, airlines in terms of their safety performance and safety culture.

The first observation I would make on the error management model is that it seems to treat safety as a static property of the system. We set up a set of rules and provided they are followed then the aircraft will remain safe. Unsafety arises when departures from the rules occur. The second point I would make is that an absence of failure does not indicate the presence of safety. Individuals will differ in terms of their basic ability, level of competence, degree of motivation, tolerance of stress and so on. Each combination of qualities and characteristics possesses its own inherent level of risk. Although the observed performance may be compliant, the extent to which the actor contributes to the burden of risk borne by the operation is masked. It is for this reason that I have argued in the previous chapter for CRM to be seen as an integral part of competent workplace performance.

In some ways we have now come full circle. When we looked at a definition of safety we saw that the idea of risk featured strongly. We have seen that attempts to understand the way safety is created in the workplace confuse the imposition of order from above with the preparedness of individuals to work a particular way. Inherent in this conflict is, again, the concept of risk. We can suppose that an individual's probability of failing to deliver the desired performance will be influenced by a host of factors. Our aim, as trainers, is to heighten the awareness of risk – the lens approach discussed in the previous chapter – while developing operational competence. Through this approach I want to place the emphasis on managing risks in the first place rather than trapping substandard performance after the event. I also said in the previous chapter that the flipside of the lens approach was the toolkit. I would argue that the toolkit analogy could possibly represent an attempt to patch up poor performance. Through better design and delivery of hard skills training, we should be able to minimise the reliance on patches. Our combined hard and soft skills training should be directed at producing effective risk managers. Of course, I have already said that risk is ephemeral, made worse by the unpredictable way events promulgate through tightly coupled systems. As we go through the rest of this book, we need to be aware of the complex interplay between the contradictory forces charted in Figure 2.2 and the probability of our training interventions having a successful outcome. In order to achieve this goal, we need an understanding of what is happening in the real world, we need a clear understanding of what we mean by safe behaviours and, finally, we need a method of dealing with departures from the desired performance. We will explore these ideas further by looking at how we build learning organisations.

**Building a Smarter Airline**

So far we have seen that safety – like CRM – is a loose, vague subject, the meaning of which varies, at the very least, according to who you are in an organisation. To make matters worse, we cannot be certain that we have a complete picture of the risks and possible outcomes in the industry. For example, historically we have compartmentalised safety within aviation into Health and Safety, on the one hand, and accident prevention, on the other. More aviation personnel are harmed every year as a result of events falling under Health and Safety regulations and yet most discussion of safety in aviation ignores this important area. The blind spot is an artefact of the division between different regulatory bodies. If we return to the BAe 146 for a moment, the question of cabin air quality is an issue for Health and Safety but investigation of the problem was delayed because the responsibility for issues within the aircraft once the doors close is vested in the country's aviation authority. The Australian experience clearly established cabin air quality as a safety issue once pilots started to suffer incapacitation in flight. From this example it is probably fair to say that we need to take a broader view of the problem than, perhaps, has been the case up to now. Just to underline my position, many safety initiatives in the construction industry have been directed at what is known as the '6 foot drop'. A fall from this height, not uncommon on building sites, is sufficient to kill and maim. However, in 1999, 75 crew and ancillary workers – maintenance, cleaners, catering – fell out of parked aircraft in the UK. In 2000, 2 cabin crew were killed when they fell from aircraft doors. Although the construction industry has recognised a significant problem and is taking action, the aviation industry does not seem to appreciate that it suffers the same problem, albeit on a smaller scale.

The idea of a learning organisation is rooted in the fact that training systems will always be inadequate in that they cannot convey the entirety of what an individual needs to know about a job. Furthermore, as circumstances change over time, skills and knowledge will become time-expired. We can set up formal communication channels designed to update employees but, in reality, knowledge tends to be communicated informally through anecdotes and 'war stories'. Learning organisations try to tap into all possible methods of generating expertise. A learning organisation displays 4 key characteristics (Easterby-Smith and Araujo, 1999):

- It actively generates information.
- It integrates that information into the organisational context.
- The organisation makes a collective attempt to interpret the new data through some form of interaction.
- Individuals and groups are encouraged to take action on the basis of a shared understanding of the problems raised.

If we apply the learning organisation model to aviation safety we can identify 3 key components:

- An accurate database of safety-related data relevant to the organisation.
- A forum for debate about issues arising from the data.
- A culture of trust that empowers individuals to apply the lessons derived from the data.

These key components, ironically, also point towards possible barriers to learning in organisations. First, as mentioned above, do we actually know what goes on in an airline in terms of unsafe practices? Second, to what extent are airlines prepared to discuss safety-related problems and, finally, what levels of empowerment are permitted within airlines? As part of this discussion of safety and organisational learning, I want to explore the components, and the barriers, in more detail.

## The Database

All aviation authorities require a minimum level of adverse event reporting and a list of mandatory reportable events is usually·published in a company manual. Over time, reporting systems have been extended to other workgroups within the company and have been broadened to include a wider range of events. The application of Total Quality Management models and Safety Management Systems are really attempts to develop performance monitoring which, of course, requires reports on sub-standard product delivery to be made. In much the same way, external audits such as the Line Operational Safety Audit (LOSA) are attempts to generate a snapshot of performance in a single report. However, there remains confusion in many airlines as to what to report and how to report. For example, look at this data for go-arounds in one airline that requires a report to be raised in the event of such action.

The data demonstrates that a significant proportion of these events were beyond the control of the flight crew; vagaries of weather and the actions of others were the problem. When we spoke to pilots in the airline we found a rather messy situation. Some pilots were of the opinion that reporting should be for abnormal events. Therefore, as a go-around is part of the briefed approach, if you actually need to execute that part of your plan, it must be a normal event and therefore does not require a report. Other pilots expressed the opinion that to require a report was, in fact, simply a way of applying subtle pressure to land at all costs. The management perspective was that the data allowed them to identify problems at airports that may require action. For example, because the company business model was based around using cheaper, out-of-town airports, the level of ATC capability was less that one might expect from busier hubs. The airspace utilisation around some of the airports was not always what you might expect and more than one approach was broken off after the pilots reported parachutists descending around the aircraft on final approach.

**Table 2.2 Go-arounds in a UK Airline** (over 12 months)

| | | |
|---|---|---|
| Total Go-arounds (12 month period) | 134 | |
| Reasons for Go-around | | |
| Weather | 43.7% | |
| Runway Blocked | 35.5% | |
| Approach Unstabilised | 11.8% | |
| Undetected tailwind | | 2.9% |
| Profile management by ATC | | 4.4% |
| Pilot handling | | 4.4% |
| Technical | 5.9% | |
| Other | 2.9% | |

Even where an event has been defined as requiring a report to be raised, the motive for the report, the definition of an abnormal event and the interpretation of an event in the light of that definition can result in variations in behaviour. The situation gets even messier when we consider the nature of the working day. We have heard many times from crew, especially cabin crew, that they see their jobs as being operational problem-solvers. If something goes wrong they simply fix it. Therefore, as they are only doing their jobs, they fail to see the need to write a report afterwards. Besides, they are often tired and in a hurry to get home.

From this discussion it is clear that the data we use to assess safety in an organisation is patchy and inaccurate. In the final section of the book we will be looking at measures of performance and changes in reporting rates is sometimes used to assess the effectiveness of CRM training. But if our reporting is not representative in the first place then what value can we place in a change in reporting?[3] From a learning organisation perspective, there are clearly issues around data gathering. We will now look at how data is used.

## The Forum

The next component of a learning organisation is some vehicle for dealing with the data collected. We need to engage with the data in order to create knowledge, a process illustrated in Figure 2.3. At the same time, we are moving from what is known as explicit information – the things we 'talk' about – to implicit, or tacit, information – those things we 'know'.

For this process to work we need to consider what happens to reports incorporated into our database. There are 2 main processes by which airlines engage with data about their operation: investigations are conducted and reports are issued. I have distinguished between the 2 because investigations do not always lead to reports and reports are not always based on investigations.

An investigation sets out to identify cause. That said, an investigation is not conducted in isolation; those who instigate the work and those delegated to conduct the work bring an agenda to the process. The net result is that an investigation is rarely an exhaustive examination of all possible causes and effects conducted in an attempt to arrive at a position of understanding. Most investigations stop at the point at which an acceptable probable cause is found (see Dekker, 2002). In part, this is because of resource limitations: thorough investigations take time and effort. As sources of information, investigations need to be treated with caution.

Reporting of information allows the lessons learned to be communicated across an organisation – at least in theory. Notwithstanding attempts to build global information sharing networks, airlines remain secretive. Therefore, releasing data – even for internal consumption – runs the risk of leakage. I was once working with Swissair and had been given some copies of their internal monthly incident report summaries. Sat on the aircraft from Zurich to London, I was reading the reports when I came across an admonition not to release the documents to people outside of the organisation. Apparently, someone had been seen on a Swissair aircraft reading a copy of a monthly summary.

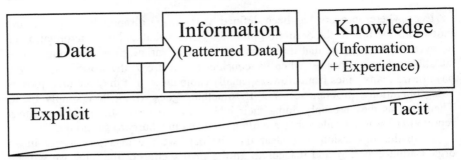

**Figure 2.3 Creating Knowledge** (after Choo et al 2000)

In an increasingly litigious world, airlines may be forgiven for being cautious about releasing data.[4] However, because reports are abbreviated descriptions of events, from an organisational learning perspective, much valuable contextual data is often lost. Many airlines use newsletters and notice boards to disseminate information, which means that communication across the whole organisation is haphazard.

The implication of this brief discussion of the investigation and reporting processes in an average airline shows is that few could be considered to meet the criteria for being considered as 'learning organisations'. Taken together, data gathering and application are conducted more in order to achieve regulatory compliance than to develop a smarter organisation. However, the final component of a learning organisation was a supportive company culture – the final barrier to learning!

**The Culture of Empowerment**

In fairness, airlines do have something of a quandary to deal with. On the one hand, standardisation is canonical in an organisation based on temporary workgroups coming together to achieve a task, many of whom may never have met one another before. To then empower the workforce to make decisions and to adapt processes may seem a contradiction. Yet the reality of line operations is that improvisation is probably more widespread than airline management would like to admit. The important point in all of this is that empowerment needs to be based upon knowledge. For the reasons discussed, barriers to learning from experience result in partial, biased, bounded transmission of knowledge, which then results in a narrow organisational response. In short, even if we did know what was going on, we probably couldn't do anything about it because of the organisational blinkers we would be wearing. We will look at organisational culture in more detail later.

## What Does this all Mean for Training?

The case I have tried to build up in this chapter goes like this: safety is the freedom from unnecessary risk but, as I have tried to show, risk is a fuzzy, messy concept. Furthermore, an organisation creates an environment in which staff act in a particular way with respect to safety. Organisations can work towards a safer environment by collecting data and mitigating the risks inherent in the workplace. Before we consider the implications of all of this for us as training developers, I want to challenge this position which, to a large degree, underpins most of how the airline industry tackles safety. The starting point for this discussion is my repeated experience of classrooms full of delegates who have never encountered anything untoward in many years of flying. Accidents and incidents are not only unknown to them, they do not have any workmates who have experienced anything more severe than a delayed departure. Everything we have been discussing in this chapter is in the realms of the abstract. If it is the fact that few of our potential trainees will have even indirect experience of the issues under discussion, what effect will any training course have on their future performance?

Cohn (2000), talking about how diabetics construct their view of the condition, makes the point that we place the individual at the centre of 'risk' and we assume that the individual then constructs defences to keep risk under control. What this interpretation means is that, first, we provide knowledge and then we assume that the individual applies rational behaviour, the result being a safe outcome. However, diabetics will act irrationally, constructing views of their condition that, to an outsider, make little sense. The information provided to the sufferer is reconstructed in the light of that individual's direct experience and not on the basis of statistical models provided by doctors.

Our earlier definition of risk needs to be developed to incorporate the means by which probabilities and severity of outcome can vary. The case developed in this chapter lacks a connection between the distribution of notable events (our database) and the concept of safety. Statistical distributions do not represent causal explanations. When we consider the shift towards Threat and Error Management (TEM), the position gets worse. The TEM argument is based on identifying possible threats to safety and then finding ways to manage those threats. At one conference I attended a speaker from a Scandinavian country expressed concern that the average number of threats his airline's operation was exposed to was higher than other operators of their aircraft type. Given that this airline flies in the arctic north, in mountainous terrain and in the arctic winter it should come as no surprise that the 'average threats per sector' exceeded that encountered by other airlines. However, returning to our analysis, the greater the number of risks present in a situation, the more events appear to contradict an individuals' own experience of that situation. An individual's past experiences will be used to develop their current subjective experience of risk which, in turn, will be used to make future predictions of risk. But the line that runs from the past to the future is not a single trace through a sequence of historical events. The 'past' is, in fact, a construction in the mind of the individual. That past is, in part, constructed through exposure to safety training programmes. And here we come against further problems. On most CRM courses delegates study accidents and incidents. Causal factors are suggested and lessons learnt. In reality, developing plausible explanations for failure helps us keep apparent chaos under control. By attributing cause we establish order. At the same time we run the risk of reinforcing the 'otherness' of failure. Because we ourselves have never had direct experience of the events being analysed, we can easily attribute the failure to some shortcomings exhibited by the actors in the event and sleep happily in the knowledge that it could never happen to us. Whereas the discussion so far in this chapter has been rooted in a model of logical behaviour, this last section seeks to introduce some of the randomness apparent in the real world.

How, then, do we take this illogical model into account when developing our courses? First, we need to establish the concept that the laws of probability tell us that we are all equally exposed to a risk. Rarity of an event within our experience does not mean that we are less-likely to encounter that event. Probability and consequence have long presented problems for the selection of appropriate case studies to use in training. Pilots, especially, are adept at dismissing the actions of colleagues involved in accidents as aberrations. Moreover, the individuals in the classroom would never be so stupid as to commit the mistakes made by the accident crew. It can very difficult – at times impossible – to get the class to identify cause as opposed to stating what the crew *should* have done. The more removed the case study in terms of seriousness, geographical location, scale of disaster and so on, the more readily some trainees can deny the lessons to be drawn from the event.

The next issue relates to the prevalence of statistical approaches to safety. Airline Safety Management Systems (SMS) gather events and assign them to cells in table establishing frequency and severity. But how does that square up against the concept of risk as being socially constructed? This is probably reflected in the difficulty I have encountered trying to get crews to identify the risks they are exposed to in the workplace and, more important, how those risks can be translated into hazards by inadequate crew performance. Hard data, in the form of accident and incident reports, cannot be denied. The problem is that much of what we mean by safety resides in 'grey data' (Falconer), by which I mean those events which may be known to the work force and which are usually worked around in some way. Numerous, apparently trivial, everyday events rarely seem to coalesce to the extent that they are perceived as threats to safety. Consider the difference of opinion about executing a Go Around discussed earlier (p24). Different pilots construct their own model of a supposedly standard manoeuvre designed to keep the aircraft safe.

It goes without saying that we need a clear understanding of safety in the workplace because we have identified CRM as a possible way to 'improve' safety. The concepts of probability and consequence seem to give us an order of magnitude for the issues we are dealing with. The concept of context, or perspective, opens up a different set of problems. The earlier discussion of context considered the different points of view of stakeholders in the airline system. In short, who says what is a risk and what is not? The answer will depend upon your point of view and whether you are directly exposed to the risk. But even then we have just seen how risk is probably constructed in our consciousness rather than existing in some absolute sense. Finally, we need to ask: whose view of safety should prevail? This is especially important if the views of training budget-holders do not coincide with the views of those on the line.

## Conclusion

It is not my intention here to discuss safety and safety management in any detail. At the end of the chapter I have recommended some authors for those readers who want a deeper understanding of the issues. My aim in this chapter is simply to raise some questions about a concept that underpins everything we are trying to do through CRM training. We stated that safety was a goal of CRM training and so it is legitimate that we try to make that goal concrete so that we can see what we are aiming at. Our immediate concern is two-fold:

- What do we need to train (and why)?
- How good is the data being used to drive our training?

To some extent, the first question has already been answered by the regulatory authorities in the various published guidance documents. We still need, though, to shape, add emphasis and to adapt to our company requirements. We want to guarantee completeness. We may also want to consider the question of added value. The second question is crucial for 2 reasons. First, if we use distorted information in training we are in danger of drawing the wrong lessons. Second, if we lack reliable data we cannot measure the effectiveness of our efforts.

## What to do Next

Think about your own airline. What reporting systems exist? How is the information used? How is information fed back to the front line? When was the last time you raised a report? When was the last time you should have raised a report and didn't? Why not? How does senior management respond to incidents on the line?

## References

ABC (2000), 'Whistle-blowing pilot still flying suspect planes', Australian Broadcasting Corporation News Report, 2 February 2000.
Choo, C., Jerell, H. and Landay, W. (2000), *Program Management 2000*, Defence Systems Management College Press, Fort Belvoir.
Cohn, S. (2000), 'Risk, Ambiguity and Loss of Control', in P. Caplan (ed.), *Risk Revisited*, Pluto Press, London.

Easterby-Smith, M. and Araujo, L. (1999), 'Organizational learning: Current Debates and Opportunities', in Easterby-Smith, M., Burgoyne, J. and Araujo, L. (eds) *Organizational Learning and the Learning Organization*, Sage, London.

Falconer, L. (2002), 'Management decision-making relating to occupational risks: The role of 'grey data', *Journal of Risk Research*, Vol. 5 (1), pp 23-33.

Gherardi, S. and Nicolini, D. (2000), 'The Organisational Learning of Safety in Communities of Practice', *Journal of Management Inquiry*, Vol. 9 (1), pp 7-18.

Hollnagel, E. (1999), *Accident Analysis and Barrier Functions*, Kjeller (Norway) Institute for Energy Technology.

Itoh, K., Andersen, H.B., Masaki, S. and Hoshino, T. (2000), 'Safety Culture of Track Maintenance Organisations and its Correlation with Accident/Incident Statistics', *Proceedings of the 20th European Annual Conference on Human Decision Making and Manual Control*, pp 139-148, Copenhagen.

Paterson T.T. (1955), *Morale in War and Work*, Max Parrish and Co., London.

Reason, J. (1990), *Human Error*, Cambridge University Press, Cambridge.

Rochlin, G.I. (1999), 'Safe operation as a social construct', *Ergonomics*, Vol. 42 (11), pp 1549-1560.

Wiegmann, D.A., Zhang, H., von Thaden, T.L., Sharma, G. and Gibbons, A.M. (2004), 'Safety Culture: An Integrative Review', *International Journal of Aviation Psychology*, Vol. 14 (2) pp 117-134.

**Additional Reading**

Beck, M. (2001), 'Learning From Disasters: An Organisational Memory Management Perspective', *Institute of Occupational Health and Safety Journal*, Vol. 5 (2) pp 21-32.

Dekker, S. (2002), *Field Guide to Human Error Investigations*, Ashgate, Aldershot.

Perrow, C. (1999), *Normal Accidents*, Princeton University Press, New Jersey.

Weick, K.E. (1995), *Sensemaking in Organisations*, Sage, California.

**Notes**

1 For an interesting example of the relationship between economic decision-making and safety outcomes see the accident of Quantas Flight QF1 at Bangkok in 1999.

2 The example of the BAe 146 also shows the need to examine the bigger picture when considering how 'unsafety' is created in organisations, the need to 'determine the determinants' I mentioned in the previous chapter.

3 I did read of one airline complaining that crews were now 'over-reporting' having been through CRM training, which raises the question of – when is enough enough?

4 See Flight International 22 March 2005 for a commentary on this problem.

Chapter 3

# Establishing the Goal – Identifying CRM Behaviour

**Introduction**

In Chapter 1 I identified a rather broad-sweeping set of challenges for CRM training but, now, we need to focus on the matter in hand – changing workplace behaviour. In the previous chapter I tried to show that safety, considered an outcome of competent performance in Chapter 1, was partly a reflection of how the individual viewed the workplace. Although CRM issues cannot be decoupled from the organisational context, the primary goal of CRM training must be to develop the soft, or social, skills required for personnel to function safely and efficiently in the workplace. I have said that aviation is a production process. I have also proposed that safety is, to a degree, behaviour in a work context. Traditional approaches to training in aviation have concentrated on the hard, technical skills of system operation and control. Now we need to broaden our model of skilled performance. In many ways, aviation is catching up with other sectors of commerce and industry where competence-based models of training have prevailed for some years. However, in order to broaden the scope of aviation training we need some way of identifying what we mean by 'soft' skills.

In this chapter we will examine some of the precursors to safe behaviour. We will look more closely at the coupling between hard and soft skills and finally, we will explore the techniques to be used in identifying CRM skills. The desired CRM behaviours constitute the goal of CRM training and, so, it is important to establish this goal before we move into the next section of the book, which deals with training design. To develop courses without a clearly defined target is, at best, an act of faith and, at worst, a waste of money.

**What Behaviour are we Talking About?**

In this section we need to try to answer the question 'what behaviour are we trying to change?' At one level the answer to this question is simple. What we see in the workplace is the application of previously-trained skills within an operational framework defined, largely, by procedures. However, at another level, we need to recognise that behaviour is under a degree of voluntary control – the individual displays an element of agency. We will look at the desired operational performance first before looking at what might then influence the actor to execute the desired behaviour. We will end this chapter by looking at some example frameworks of behaviour we might want to promote through training.

*Separating Hard and Soft*

In a study that tried to link the probability of a pilot experiencing an accident with that pilot's recorded performance in training, 3 clusters of behaviours emerged (Feggetter, 1990). The first dealt with the pilots' psychomotor skills. In this cluster we can see such things as reaction times, recall of information, division of attention, accuracy, control and coordination. Many of these headings would form part of a lesson on Human Information Processing. The next cluster related to what the author labelled 'secondary task/secondary skills' and comprised managing radios, planning, navigation and map reading. Finally, the author identified a task effectiveness cluster that included attitude and captaincy. The 3 clusters identified sit on a spectrum from the conventional aircraft control skills at one end to a subset of what we would consider to be CRM skills at the other. It is important to note that the study looked at military helicopter pilot training records. If you also bear in mind the date of the study, it is still interesting to note that what we would now label as CRM skills were readily apparent and significant. The idea of clusters of skills applies equally well to other aviation groups. For example, cabin crew demonstrate behaviours in 3 key areas: delivery of service, management of people and management of safety. Within each cluster it is possible to identify some specific technical skills and some generic social skills. For maintenance we can propose clusters around system diagnostic testing, replenishment and replacement, procedural activity to sustain airworthiness and so on. To explore this idea further we are now going to look at flight deck crew at work. I want to look at the flight deck from 2 perspectives: first, we will explore an accident that occurred in 1999. Then we will look at some pilots discussing a particular in-flight hazard – flight in icing conditions.

**Tragedy at Little Rock**

On 1 June 1999, American Airlines Flight 1420 was approaching the Little Rock airport from the south, the MD82 crew having been advised by ATC of bad weather. The crew could see heavy thunderstorms moving in from the northwest. Although the intention is to fly an Instrument Landing System (ILS) approach to runway 22 Left, the Approach Controller asks the crew if they can see the runway and suggests that they try a visual approach instead, which the crew decline. Now the winds shift as the weather front tracks closer to the airport and the crew are faced with a possible tailwind landing on a wet runway. They change to runway 04. Too close to the overhead, the crew have to turn outbound to position for their new approach. With conditions deteriorating, the crew discuss with the Controller the possibility of making a visual approach to get in quickly. Having received clearance to land the crew lose sight of the field and have to revert to their planned ILS. As they finally get established on the approach the storm arrives over the airfield. In poor visibility, the crew are off course and having difficulty locating the runway. 'This is a can of worms' comments the captain. The aircraft touched down at 2350 local time and ran off the end of the runway, killing the Captain and 10 passengers. The entire approach sequence took 16 minutes and 30 seconds.

## The Structure of Proficient Performance

The crew of Flight 1420 were coming to the end of a long duty day. It was nearly midnight and the weather conditions were severe, factors that were significant in terms of the outcome. However, to throw light on the concept of pilot competence, we want to examine what the crew were trying to do during the final moments of the flight. The short answer is, of course, 'to land'. First, though, let's back up a bit. It is possible to describe a trajectory followed by the aircraft from when it pushes back off the stand and ending when the parking brakes are applied at the destination. Along that trajectory the aircraft has to achieve specific configurations that we will call 'Goal States'. A goal state is defined in terms of a set of values. So, in order to land, we need a runway pointing in the correct direction. The runway must be available for use and the wind needs to be within certain limits. The aircraft must be configured correctly and our speed and rate of descent need to be within bounds. Hopefully it will also be the desired airfield. This list is not exhaustive but it illustrates the point. Immediately we can see that failure to achieve all of the parameters will result in a risk of an adverse event on touchdown. Equally, a change in certain values, such as the wind or visibility, may render the goal illegal and so we will need a fall-back plan.

Goal states are arranged along the trajectory in a sequence. Some goals are tightly coupled in that it is difficult to move onto a new goal unless the preceding goal has been achieved. For example, landing without having first met the stabilised approach parameters is not always possible. Other linkages are more fluid, such us direct routings within a flight plan. Of course, there is some flexibility within this model in that goal states are achieved within a volume of space and time. Crews can control the activity required to achieve the goal state. They can delay an action or bring it forward. They can discard a procedure if appropriate. They can even do something not prescribed in any manual. The crew's freedom of action is bounded by aircraft performance parameters, airspace regulations, company procedures and applicable law.

The technical skills of piloting an aircraft are all to do with configuration and control. So, the underpinning systems knowledge, the company procedures and checklists, the manipulation of automatic flight controls and stick-and-rudder skills are all directed at this process of successfully transitioning from one goal to another. However, the soft skills – what we call CRM – are to do with management of that process. CRM skills allow the crew to judge the rate of progress towards the desired goal, detect deviations from the desired trajectory, initiate recovery action, develop alternative plans and so on. When looked at in this context, the traditional separation of CRM from technical proficiency seems fundamentally flawed.

*Crew Processes at Little Rock*

The crew of Flight 1420 considered 5 different approaches along the trajectory towards touchdown. The initial ILS to RW22L, the Approach Controller's suggested visual approach, ILS RW04R, an expedited visual to 04R and, finally, vectors back to the ILS. We can analyse crew performance in terms of how they identified their goal, their application of rules to the process, their aircraft configuration and their management of the transitioning between goals. The evidence I use in this analysis is contained in the Cockpit Voice Recorder (CVR) transcript. From the initial contact with Approach to final touchdown, the crew make about 210 utterances, the Controller makes a further 38. There are 2 GPWS 'Sink Rate' warnings on final approach and 1 call to the Cabin.

At no stage did the crew explicitly formulate a plan for each of the revised approaches. Assuming that a briefing for the initial approach was conducted prior to the CVR transcript, only one other briefing, albeit abbreviated, was attempted. We can classify the communications types in terms of their purpose. Broadly speaking, within my analysis, communication serves 3 purposes: to clarify the procedure so that all team members are aware of intentions; to control activity within the procedure, for example, calling for flaps; to verify the status of the procedure, by which we mean confirmation and checking. An example of verifying status would be the crew discussion around crosswind limitations, the cause of the first change of plan. On the approach to Little Rock 16.6% of communication was to clarify procedures, 13.1% was to control activity and 14.9% was to verify status. Fully 17.9% of communication events were simple responses to statements, the 'yeps' and 'uh-huhs' that glue conversations together. One third of the communication within the crew was aimed at locating the airfield. Of course, the specific communication events we see within teams will vary across situations and it would be wrong to assume that a 'perfect' distribution of communication events exists. Our objective in looking at the goals of communication is to get closer to the skills we are trying to develop in proficient crews. In this case we can broadly define 4 functions being applied to the work process:

- Application of procedures.
- Control of activity.
- Verification of current status (and here we could add establishing spatial relationships).
- Sustaining social cohesion.

It is, perhaps, significant that the smallest category of communication acts was that relating to control and the one thing the crew significantly failed to do was exercise control as their situation worsened. The attention of the crew was drawn to other things, such as trying to locate the airfield. The approach developed a momentum of its own as, first, one then another of their plans had to be changed. The crew's internal processes got out of step with the demands of the situation.

As well as the type of activity observed, we can examine the effectiveness of the performance. In a qualitative sense, what is most apparent is that the crew were functioning in an implicit mode. Sentences are half finished, actual meaning hangs in the space between conversational elements. For example, as the crew are vectored outbound for the final ILS the weather is moving ever closer to the airfield overhead. The crew seem to be aware of the implications of the changes in relative positions of the aircraft and the thunderstorm cells and here is how they address the problem:

> Captain: "I hate droning around visual at night in weather without having some clue where I am".
> First Officer: "Yeah but, the longer we go out here the ...".
> Captain: "Yeah, I know".
> First Officer: "See how we're going right into this crap".
> Captain: "Right".

At this point the First Officer, the non-handling pilot, makes an unprompted call to Approach suggesting a closer turn in to the localiser.

In the discussion so far we have taken something of a top-down view of competence in that we have looked at performance within a situation bounded by the structure of the aviation task. I have proposed a cognitive model rooted in the mental control of the process and we have just explored how mental processes are manifested in crew behaviour. The case study we used is notable in that it offered several examples of redefining the target goal state but it was also apparent that the crew loosely applied control of the process they were responsible for. However, the mental processing required to support the hypothesised model of control requires crew to apply well-integrated knowledge structures. In order to explore this aspect we will now consider a different sector of aviation, that of single pilot operations.

## 'Stay the Hell out of Ice!' – The Natural History of Expertise

Over a 15-year period from 1987, Caravan pilots were involved in 113 recorded incidents, of which 26 (23%) involved encounters with ice, frost and snow. Of the icing events, 10 (38.4%) proved fatal.[1] Of the other 87 events, 7 (8%) proved fatal. Icing conditions, then, represent a serious threat to safe operations. Many of the fatalities, moreover, were the result of the pilot failing to interpret and apply information within our proposed goal state model. To explore expertise, it might be useful to know how pilots acquire and interpret the information needed to work safely and efficiently.

All the pilots in the sample held a professional license and, so, have studied Meteorology at some point. They will know about icing as a phenomenon. In the USA, where most of these accidents occurred, the FAA has published various advisory and regulatory documents. Some operators have developed 'Winter Operations' procedures. Even with all of this information, aircraft continue to crash. A clue to the problem was found on an internet forum for Caravan pilots (Anon). The thread started with a simple request for help:

> A lot of you guys have mentioned to be careful with icing or even staying out of it completely. I don't think staying out of the icing is very practical, especially in hauling cargo where everything is by the minute. Why the strong concern about icing? Are you talking about severe ice or is it that the boots are not very effective.

Here we have a pilot, apparently yet to encounter icing, seeking help. The request contains a number of distinct elements:

- Our enquirer wants to know the scale of the problem posed by icing.
- It is acknowledged that commercial constraints will influence decisions.
- He wants to know if it is something about the nature of the icing or is it the aircraft's systems that are the problem?

The 22 responses posted on the forum allow us to look closer at the nature of expertise. What is immediately striking is that only 1 posting dealt with pre-flight activity (4.5% of postings) despite the fact that the majority of incidents (58%) started with inefficient pre-flight preparation. Only 42% of events (50% of fatalities) were the result of in-flight icing while 95.5% of the postings dealt with this condition. Responses fell into 4 categories. First, we see comments relating to the design of the aircraft and its protection systems:

> 'Lots of unprotected areas, pod, landing light area, radar pod, spinner, nose bowl, most of the windshield – it gets ugly in ice pretty quickly'.
> 'Limited ability to go fast and climb fast'.

Explicit in these comments is knowledge about how the aircraft design and performance compounds the problem of icing. However, implicitly we see knowledge relating to the constraints on managing the problem – inadequate horizontal and vertical speed which, in turn, affects the ability to climb above the icing layer after the onset of ice accretion or to use speed, in conjunction with the de-ice boots, to remove the ice build-up.

The next group of comments dealt with the effects of ice on performance:

> 'It stalled without warning, no stall horn'.
> 'After about 3 boot actuations you are now at 115 kias, down from 145…and descending if you like it or not'.

The third group contained advice on avoiding icing in the first place:

> 'Know where the ice is and plan to get up high before you get to it'.
> 'Know where airports are beneath you'.
> 'Get on top and don't fly parallel to a weather front'.

Clearly, although the aircraft is cleared for flight in icing conditions, the consensus seems to be that it is best to plan to avoid icing, if possible. That said, we also see the acceptance of commercial realities. The original questioner referred to the fact that staying out of icing was not practical for cargo operators 'where everything is by the minute'. Another respondent commented that 'we don't have the time to go around'. So, we need to balance safety against practicality.

The final cluster of comments dealt with what do if you suffer icing:

> 'Don't just dump flap, ease them in'.
> 'Hand-fly in ice, make turns shallow'.

Now we are back to the reality of controlling an aircraft in risky conditions. The skill is sustaining safe flight, the knowledge is demonstrated in the adaptation of skills to suit new contingencies.

This book is about *crew* resource management and, yet, we have just been looking at single pilot operations. However, if you compare the experiences of pilots who encountered icing with the performance of the Little Rock crew, you can see that much of what we consider to be proficient aircraft operation requires effective planning, anticipation, problem solving and execution of contingencies. These are the same skills we have called CRM in multi-crew aircraft. I make the observation to reinforce my contention that CRM is not an entity in its own right but, rather, is the context within which task-specific activities occur.

## The Expertise Matrix

I embarked on this diversion into what pilots do in an attempt, first, to reinforce the inter-relationship between hard skills of aircraft control and system manipulation and the soft skills of what we now call CRM. My second aim was start to map out the territory we need to explore if we are to establish useful training goals for our CRM interventions. In Table 3.1 I have tried to represent the relationship between hard and soft skills in a matrix. On the vertical axis I show behaviour in terms of how long it takes trained pilots to recall information from memory. I have used the conventional skill/rule/knowledge framework as adequate for our purposes. Skills, being automatic behavioural responses, are generally retrieved from memory a lot faster than stored information held in memory structures. Along the horizontal axis I have categorised behaviour in terms of the timeframe over which pilots can expect to see the outcome of their actions. The first category – system – covers behaviours used to configure and control technology, the outcome of which is almost instantaneous. For example, selecting flaps or lowering the undercarriage would fall into this category. Manoeuvre refers to the response of the aircraft over the short term, such as executing a descending turn. Trajectory refers to behaviour that will have an impact at some time in the, relatively distant, future. For example, the Little Rock crew, as they approached the airport, were aware of the relative position of the weather front, its relative rate of movement and the direction of travel. By projecting forward in time they could have anticipated that they would arrive over airport at much the same time as the worst of the weather. A discrete behavioural act may contain elements that fall within more than one cell in the matrix. In the diagram, I have tried to show how both Flight 1420 and Cessna Icing (in italics) can be accommodated in the matrix.

**Table 3.1 Expertise Matrix**

| | | | |
|---|---|---|---|
| Knowledge | *'Limited ability to go fast and climb fast'.* | | *'Get on top and don't fly parallel to a weather front'.*<br><br>1420 – Relative positions of storm cells and aircraft. |
| Rule | | *'It stalled without warning, no stall horn'.*<br><br>1420 – visibility on visual approach. Use of checklists. | *Effect of increased all-up weight on ability to arrest descent.*<br>1420 – effect of tailwind and braking action on initial planned approach |
| Skill | *'Don't just dump flap, ease them in'.* | *'Hand-fly in ice, make turns shallow'.* | |
| | System | Manoeuvre | Trajectory |

As we move from the bottom left cell (Skill-System) to the top right cell of the matrix (Knowledge-Trajectory), we are moving from the traditional focus of hard skill pilot training to an emphasis on higher mental processes used to control activity in the future – something we might call Situational Awareness on a CRM course. Equally, from bottom left to top right we are also able to envisage a shift from processes that occur internal to the pilot to activities that occur in collaboration with others. In the case of single pilot operations, 'others' may refer to off-board agencies that are able to lend support in order to solve problems. In the case of multi-crew aircraft, 'others' refers to other team members on-board as well as off. However, the key point I want to make is that we should not consider soft skills in isolation. We act in an operational context and soft skills support the accomplishment of safe and efficient flight. It is probably fair to say that the criticisms of early attempts at CRM training reflect, on the one hand, the disconnection between the methods adopted in class and the needs of the students and, on the other hand, the separation of hard and soft skill domains. Together, these actions almost guaranteed that there would be poor transfer from the classroom to the real world.

Although I have used the work of pilots to explore some ways of investigating the skills we need to develop in training, the approach would be equally applicable to any of the teams that form part of the complex system we call aviation. In the final section of this chapter we will look at some specific techniques for getting closer to the soft skills we have been exploring. Before that, however, we need to examine why people do not do what we want them to when they come to work. I want to finish off the discussion we started in the last chapter and look at the concept of 'culture' in more detail.

## Discretionary Control of Behaviour

The last 10 years have seen an explosion of interest in culture in aviation. National differences have been linked to issues such as the applicability of CRM and the use of flight deck automation. Implicit in the discussion is a possible correlation between culture and regional accident rates. At the same time, and as we discussed in Chapter 1, the role of the organization as a factor in accident causation has been accepted. As we learn more about how systems fail, we seem to be creating something of a paradox. On the one hand we have a variety of, sometimes competing, theoretical models being offered to explain adverse events and yet, on the other hand, none of the theories seem to be able to explain the reality of life on the line. Given the global, and increasingly multinational, nature of aviation, what is the interplay between cultural and organizational factors? And how does all of this influence safety? If we are going to influence behaviour through training, we need to understand the context within which that behaviour is going to be displayed. I mentioned earlier that individuals exercise agency – they posses a degree of voluntary control over their behaviour. They can chose the degree to which they action a routine and they can chose to withhold their effort if they so desire. In this next section I want to take an initial, stumbling, step towards an integrated model of organizations, culture and safety in the workplace. I have tried to capture the problem in Figure 3.1. The diagram is an attempt to visualise what we can call 'discretionary behaviour'. Imagine a 'line in the sand'. This is the standard of performance we expect from all employees. It is not the maximum attainable performance in the specific area, simply the minimum standard we expect. If we meet that standard, then we know that the task will be accomplished in a safe and efficient manner. Over time we all find ways to complete a task at a reduced level of effort. We learn the various shortcuts we can take or techniques that save time and effort. We also learn that the job can be completed without necessarily following the procedure in full. Nothing goes wrong and we haven't necessarily broken any rules, we have simply completed only those elements of the procedure essential for success. This is the zone of tolerance shown in the diagram. Of course, the 'Zone of Tolerance' is a concept, not a physical entity. It is a reflection of the flexibility or resilience inherent in most systems. We cannot measure the 'flexibility' in the system in any quantitative way. We just know that, in the real world, this flexibility exists. More often than not, we see the system's flexibility when minor things go wrong but the system trundles on quite happily.

Once performance falls too short of the mark, such that we are now in the Zone of Non-compliance, then things become 'observable' and action is taken. That said, we know from the last chapter that not all non-compliance is captured within systems.

The problem with working in the 'Zone of Tolerance' is that we do not know the extent to which the risks present in the system are amplified because we are working in a sub-optimal manner. We do not know if, by taking a shortcut, we have possibly created a situation that could become a problem in the future. Consider the behaviour of the unknown personnel who used tap water instead of deionised water in our exploding water heater example. Were they simply cutting corners or were they deliberately putting aircraft at risk?

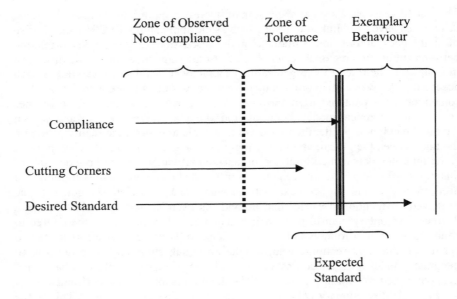

**Figure 3.1 Behavioural Control**

The final zone I have identified tries to reflect the fact that organisations want personnel to go beyond the minimum. Many people take pride in a job well done and striving for perfection can be very motivating. In that sense, the Zone of Exemplary Behaviour captures personal desires as well as organisational demands.

In training, we aim to equip personnel with the skills needed to exercise compliant behaviour. We also want to prepare people to deal with probable process failures. What we rarely do is consider the role of agency and the factors that will influence the degree to which people 'want' to comply or exceed expectations. The closest we come to considering this issue is when we attempt to characterise an organisation in terms of its 'culture', a blunt tool and probably useless for our purposes. When we looked at the concept of safety culture earlier, some of the characteristics identified reflect voluntary control of behaviour. For example:

> Safety culture emphasizes the contribution from everyone at every level of an organization.
> The safety culture of an organization has an impact on its members' behaviour at work.

Although couched in indirect terms, these statements nonetheless imply that behaviour will vary according to some apparent desire or voluntary preparedness to engage with the process of work. The model proposes that external forces shape workplace behaviour. In this case, forces flowing from this thing called 'safety culture'. Two factors most associated with non-compliant behaviour are our perceived probability of failure and our perceived probability of punishment. If we take car drivers exceeding speed limits as a classic example, the willingness to break speed limits is linked to our belief in the safety of our actions (we will not have an accident just because we are going faster than the speed limit) and our belief in the probability of getting caught (very few speeding offences – if any – in relation to the number of times I break the speed limits). The same principles underpin the willingness of personnel to comply fully with procedures. So, after experiencing a fatal accident because of improper aircraft loading, one cargo carrier changed procedures so that both the Captain and the First Officer must now sign the load sheet to verify that they have both checked the load before departure. Because of the short time between getting the final paperwork and the need to depart, this procedure was almost universally ignored. Both parties sign the paperwork but both parties certainly do not check the load. Provided that managers do not observe the crews, and that a further load-related event does not occur, the procedure will continue to be ignored.

Thinking back to our discussion of risk, modifying our behaviour on the basis of an unknown future probability – after all, what hazardous situations could we encounter while speeding on our car journey – does not seem rational and here we come back to the heart of the matter. As we discussed earlier when we looked at risk, humans do not act in a rational manner at all times. Observed behaviour in the workplace is influenced by a host of factors. We have just looked at two – the probability of failure and retribution. What is more, I have suggested that these forces are highly variable in the extent to which they shape behaviour. I now want to look at a cluster of forces we lump together under the heading of organisational culture and see if these are any more predictable in their effect.

## Why Stuff Happens

Unfortunately, even after a century of usage, no reliable definition of culture is available. It has been described as 'one of the two or three most difficult words in the English language'.  Most researchers view culture as a collection of beliefs, values, symbols, and rituals.  In aviation, current thinking is dominated by the work of the Dutchman Geert Hofstede who, in the 1970s, was employed by IBM to investigate differences in the company's operations around the world (Hofstede).  The project was enormous, with employees completing over 100,000 questionnaires.  The resulting analysis produced a 4-factor model: Hofstede reckoned that measures of Power Distance, Ambiguity Avoidance, Masculinity and Individualism were enough to sum up nations.  Researchers in Texas gave weight to the model when similar national distributions were found in responses to a pilot attitude questionnaire (Helmreich & Merritt).  Unfortunately, the model has done little to bring clarity to the matter.  It seems that culture is everywhere you look.  The Texas researchers have identified national, organisational and professional variations and, of course, we must remember safety culture as well.  As we saw in the last chapter, other work has shown that there are as many cultures in an organisation as there are discrete work groups.  Culture, it seems, is not a constant.  It exists in different flavours and can change over time.

*Conceptualizing Culture*

Culture theorists fall into 2 camps.  One group sees culture as an object while the other treats culture as a process.  The 'culture-as-object' school – Hofstede's followers – view culture as a property possessed in some way by group members.  The problem with this approach is that, so far, we have yet to isolate the constituents of the property in such a way that we can exercise control.  The 'process' view is that human interactions are the result of cognitive activity.  Whenever we interact with another person we are constantly assessing progress, making sense of what has happened and deciding what to do next, gauging success and deciding on the bounds of our actions.  The 'culture-as-process' school would argue that what we call 'culture' is created between individuals and only exists to the extent that we recognise that our interactions fall into patterns.  These patterns are shaped by the past history of the group and by the forces that constrain our freedom of action in the present.  The important point in all of this is that the 'ingredients' of 'culture' represent a further group of factors that will influence our preparedness to act according to some preferred model of performance.

    If we follow the 'culture-as-process' approach, we see that culture is the result of the myriad of negotiations we enter into as part of the working day.  For example, imagine that the cabin crew on an aircraft during a turn around will have to contend with issues to do with cleaning the cabin, replenishment of the catering supplies, perhaps having to deal with maintenance because of a technical problem and negotiating with gate agents and pilots over boarding times.  In addition, they will have procedural activity associated with safety and security to complete – and they might even want to take a break for a cup of tea.  Managing all of this activity requires negotiation between agencies and, probably, within crewmembers.  If we also follow the line of argument that says that people exercise control over their behaviour, then we can say that the extent to which our cabin crew will be proficient – that is, safe and effective – will vary according to how they view the context of their work; their view of the company culture, in effect. I want to explore this idea of a negotiated workplace further in the next section.

## The Life of Organisations

Organisations work at 3 levels. First, we have strategic management. Here, the corporate business model is created and translated into policies and procedures. These are then handed to the next layer in the model, line management. This layer controls the daily conduct of the business. The final layer comprises the workforce, the people who turn procedures and assets into productivity. The relationship between the 3 layers is illustrated in Figure 3.2.

The roots of organizational culture lie in the ebb and flow across the management/workforce interface. We can identify 4 key areas of interaction: effort, the style of management, autonomy and, finally, the ambiguities endemic in the workplace (Hodson). In each of these key areas line management make demands of the workforce and, in return, employees develop strategies for dealing with those demands. For example, in the case of 'effort', the contract of employment usually specifies a minimum number of hours to be worked in return for the agreed reward. However, in reality, effort demanded often exceeds the level initially agreed and forms of compensation – bonus payments, overtime, time off in lieu – exist to provide for this additional contribution. Expected demand is reflected in the contract and the required level set can sometimes be excessive. Some airlines use the legal maximum duty hours as a planning target. However, in certain types of operation, such an intensity of flying is not sustainable by the individual; fatigue begins to take its toll. Recruitment issues play a part. Some airlines do not employ sufficient staff to meet the task relying, instead, on overtime to make up the difference. In reply, workers employ a range of strategies to control the perceived level of overwork. They refuse to work overtime, report late, report sick and, in extreme cases, leave to join other airlines.

Style of management is self-explanatory. Few airlines invest in the training of people appointed to be managers. In some cases, airlines simply fail to employ sufficient managers. As a result, the management process can be chaotic and abusive. Workers respond through ridicule or negation of the management role. In some cases, alternative systems of informal management develop as a way of compensating for inadequate formal management – people 'get on and do it'.

Autonomy reflects the fact that workers seek fulfilment through responsibility for their actions – what Maslow called 'self-actualisation'. Management delegates a degree of control to workers but problems arise when responsibility and authority are not aligned. For example, we have created the stabilized approach point as a safety measure and, yet, many pilots know that to execute a go-around may attract unwanted attention from their fleet manager. Again, the Captain has the discretion to extend the crew duty day. Captains who fail to exercise that discretion to the advantage of the company can have their judgment called into question.

The ambiguities of the workplace include such initiatives as Total Quality Management (TQM), safety reporting, the routine use of recorded flight data in fleet management and so on. At face value, these strategies seem to be based on a rational approach to management. In reality, they can translate into increased workload and reduced responsibility for the workforce. For example, as we saw earlier, an airline can require an occurrence report to be raised in the event that the pilot makes a go-around. From the airlines' perspective, it wants data to control operational risks. From the crew perspective, it is just one more way for the airline to exact punishment.

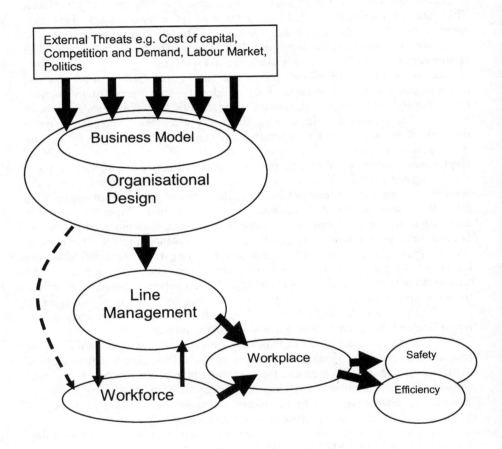

**Figure 3.2 How 'Culture' Gets Made**

These examples are typical of life in the real world of airline management. An airline will develop procedures that allow it to exercise a satisfactory level of control. Those procedures represent the aggregated knowledge in the company about efficient aircraft operations and workforce management. Airlines will differ in the effectiveness of their total suite of procedures. In reality, what we see is the importance placed on managing the technology in airlines, often at the expense of managing the people.

Tensions between the workforce and line management are resolved in the workplace. However, aviation has an extra problem in that direct control of the workforce is exercised at arm's length. Crews report for duty and immediately disappear for the whole working day. On return to base, the crews disperse as fast as possible. This is part of the challenge of airline management. The efficiency with which crews complete their tasks will vary according to their aptitude, level of competence, physiological and emotional state at the time. The preparedness of the workforce to sustain the highest level of engagement with the task will be a reflection of the culture created by the management/workforce dynamic. In turn, safety can be seen as the level of risk considered tolerable by the workforce from moment to moment. In part, safety is linked to efficiency in that inefficient working can induce an increase risk of process failure. At the same time, because we have no instantaneous measure of the level of risk present within the work process, airline employees are not able to determine the incremental risk induced by a sub-optimal, but still possibly acceptable, task performance. Let's go back to stabilized approaches. Although airlines define a fixed point on the approach, pilots are clearly prepared to extend the volume of space within which they achieve the stabilized approach parameters, often to the point of touchdown. In fact, a stabilized approach is more a reflection of each pilots' perception of the risks inherent in their approach management techniques. The number of runway excursions after landing resulting from hot and high approaches is evidence that risky behaviour exists. Of course, we have no data for how many approaches are technically unstable but remain tolerable to the pilots on the flight deck and do not result in the need for reporting action after landing. In these cases, 'safety' is a mental – or social – construct generated afresh on every approach and, in all probability, influenced by factors off the flight deck. If this line of reasoning is valid, then what we call 'safety' really is a product of the workplace and not a precursor.

We have taken this diversion through the esoteric world of voluntary control of behaviour in order to cast into harsh relief some of the issues we will face when designing behavioural modification courses: CRM classes, that is. The problem will be all the more apparent when we consider how to measure the effects of training, which we will do in the final section of the book. For now, I want to return to the matter in hand; what is CRM behaviour.

## So – Back to Behaviour

I have laboured the point that I take as a guiding principle of CRM that technical skills (manipulation of controls and devices) need to be integrated into non-technical dimensions of performance (social or soft-skills). The technical skills can be measured in terms of adherence to procedures, correct configuration of systems and devices and, of course, accuracy and smoothness of aircraft handling. The non-technical skills pose different problems. For a start, much of what falls within the domain of CRM occurs between one or more people (if you are confused by that statement, then consider a single pilot talking to themselves as things start to go wrong or the brainpan chatter all of us experience everyday as we go about our work – can we not have interaction with ourselves?). There are multiple players in this game. Whereas, in the case of technical skills, one of the players (the aircraft) is usually fairly constant for all those being assessed, in the case of the non-technical skills, no single variable is held constant. The performance of the subject being assessed is intimately associated with that of the other players in the situation. The fluid nature of the crew process will have implications for assessment and we will deal with these in the final section of the book. For now, though, what are these soft skills we have been talking about for so long? To answer this question it may be useful to review previous attempts to isolate effective crew behaviour.

One of the first structured attempts to capture behaviour was the development of the Crew Effectiveness Markers by NASA and the University of Texas. The research team analysed a selection of accidents cause primarily by poor human interaction. The NASA/UT Crew Effectiveness Markers comprise 13 categories of behaviour. I have taken just 3 here as examples:

LEADERSHIP
        Captain showed leadership and coordinated flight deck activities
            In command, decisive, and encouraged crew participation
ASSERTIVENESS
        Crew members stated critical information and/or solutions with appropriate persistence
            Crew members spoke up without hesitation
WORKLOAD MANAGEMENT
        Operational tasks were prioritised and properly managed to handle primary flight duties
            Avoided task fixation
            Did not allow work overload

The formal introduction of CRM as a training requirement within the JAA area of influence also introduced an intention to assess soft skills at some future time. To achieve that goal, a template marker scheme was developed. The NOTECH/JARTEL project has developed marking schemes and methodologies for crew assessment and instructor standardisation. Based on an aggregation of various schemes in use in European airlines, here is how the NOTECHS marker scheme describes a comparable set of behaviours to those illustrated above:

Category: LEADERSHIP AND MANAGERIAL SKILLS
      Element: Use of authority and assertiveness
           Example: Takes initiative to ensure involvement and task completion
      Element: Maintaining standards
           Example: Intervenes if task completion deviates from standards
      Element: Planning and coordinating
           Example: Clearly states intentions and goals
      Element: Workload management
           Example: Allocates enough time to complete tasks

In this scheme, CRM behaviours fall into one of 4 categories, each category comprising a number of elements. Examples are offered to elaborate on each element. The first thing to notice is that there seems to be several ways to categorise behaviour. As we look at further examples you will notice that marker schemes seem to comprise ad hoc collections of abstract statements, descriptions of observable behaviour and a sprinkling of wishful thinking.

At this point it would be useful to establish the 2 distinct functions of a behavioural marker scheme. On the one hand, we want a benchmark against which performance can be measured and, on the other hand, we want guidance as to what skills need to be developed in training. These 2 different outcomes impose constraints on how we develop our markers. The 2 schemes we have looked at tend to fall more into the first category in that they set out to establish the desired performance. Individual performance can be viewed against the benchmark and shortfalls can be identified.

Observational schemes, like any form of test, have a requirement to comply with the principles of validity and reliability. Validity is the degree to which the measurement tool actually measures the property under investigation and reliability is the degree of consistency of the measurement tool over time. So, if we look at the NOTECHS example, if I observe a pilot and note down all the examples of behaviour that match the elements within the category, am I actually sampling 'Leadership and Management Skills' or am I measuring some other cluster of behaviours but they just happen to fit this description? The extent to which the behaviours are, indeed, leadership and management would be a measure of validity. If then I observed that pilot for a second time, say 12 months later, the degree to which I assign the same behaviours to the same category (or assigned the same grade to the same performance) would be a measure of the reliability of the tool. This complex topic will be discussed in more detail in Section 3 of the book but for now we need to understand that the needs of validity and reliability put constraints on the way we describe behaviour and these constraints can run counter to our needs when it comes to course development.

To illustrate this I now want to look at an example of a behavioural marker based on fieldwork I did with a US regional airline. Again, I have selected a set of markers that represent the same cluster of behaviour we looked at earlier:

TASK MANAGEMENT

This dimension relates to the conduct of the task. It includes the consistent and appropriate use of checklists and procedures. Making effective use of time. The avoidance of distraction and maintaining the bigger picture of things happening around the aircraft.

Positive indicators include:
A consistent, but flexible, use of SOPs. Monitoring the use of checklists during busy periods and the positive verification that tasks have been completed. Maintaining an even tempo of work (no unnecessary haste or urgency). Recognising when to minimise non-essential conversation.

Maintaining awareness of other aircraft, objects etc around the aircraft both in the air and on the ground. Actively developing mental pictures of what to expect during the next stage of flight (e.g. through verbalisation of expected landmarks, events, system changes etc). Anticipation and thinking ahead. Being aware of time available/remaining, being aware of things around the aircraft (in the air and on the ground), verifying geographical position.

Negative indicators include:
Too strict an adherence to or rigid application of SOPs. Spending too much time out-of-the-loop on admin tasks, failure to update on events when off-frequency. Rushing or delaying actions unnecessarily

TEAM BUILDING

This dimension describes the extent to which effective working relationships are established and maintained within the crew. It includes behaviour which binds the team and which establishes a task focus.

Positive indicators include:
Setting the tone. Clarifying expectations and standards of performance. The recognition that others have a part to play in the crew process. Clear allocation of tasks and responsibilities. Briefing any excursions from SOPs. Fostering a sense of comfort and inclusiveness in the group.

Negative indicators include:
Avoiding responsibility for actions, preventing full expression of views, intolerance, failure to allow individuals to fulfil their role, interference in the work of others.

This third set of markers was developed by interviewing line pilots using the techniques of Critical Incident analysis and Repertory Grid, which I will briefly explain next.

*Critical Incident Technique*

The Critical Incident method of conducting a structured interview requires the interviewee to identify exemplars of good and bad performance. It is important to stress that the examples must be from their direct experience, not what they have heard from others. Ask them to explain what happened, why it happened, what the various actors were doing and how successful they were. Ask why, in their view, the incident was important. As a basic rule, an interview will prove more valuable if, for each event, we can get the interviewee to describe situations, task, actions and results. By this we mean what was the background to the event, what was supposed to happen, what actually happened and what was the outcome. We are trying to identify examples of good and bad practice in order to establish the standard of performance expected and the skills to be employed in the under investigation. Try to get a range of examples relating to the same aspect of the job.

It is often easier to identify bad performance than good performance but you should try to get exemplars of perfection as well as disasters. At the end of this exercise you should have a set of examples, which is always useful for when we come to lesson development, but more important, you should have a set of behaviours.

## Repertory Grid Technique

The Repertory Grid process can be time-consuming and difficult to use so I offer here a shorthand version, so to speak. It requires the interviewee to identify a range of people whose work-place performance they are familiar with. They need to write some form of identification mark for each of the people on a separate piece of card. It is important to stress that the interviewer is not interested in who these people may be; we will simply be using them for comparison. You will need a range of performers, some good and some not so good. Shuffle the cards and extract three. Place two together on one side and put the third off to the other side. Ask the candidate to think of the three people represented by the cards and get them to try to think of a quality or characteristic, in relation to the task being investigated, which the pair share and which separates them from the singleton card. Once the distinguishing feature has been identified, ask for clarification and, if possible, an example of the behaviour in practice. As an alternative, you can allow the interviewee to decide, from the 3 dealt cards, which 2 have the strongest association.

It is the comparison task that can cause confusion. For example, because the cards are selected at random, it is conceivable that we will have a good and bad performer as the pair with another good performer as the singleton. The tendency is to compare good with bad and so the interviewee will mentally re-combine the selection in order to isolate the poor performer. You need to be on your guard to ensure that the method is applied as intended. It may well be that the interviewee cannot identify any distinguishing characteristic shared by the pair. If so, then simply reshuffle and try again. Keep working through the pack of cards, pulling out, at random, three cards at a time, until no further information is obtained.

A shortened method we have found to be useful involves getting the interviewee to identify a range of people, between six and ten, and to write their names or some other identifier on bits of card. Shuffle the cards and then draw out pairs. Get the interviewee to compare the two people named on the cards in terms of the aspect of the job under consideration. For example, you may say that you are interested in who would make a good senior cabin crew member. The interviewee draws up a list of senior cabin crew whose performance is known to them. As each pair is drawn out, we ask the interviewee to identify differences between the two individuals in terms of how they do the job. By repeating the pair-wise comparisons, we can draw up a list of effective and less-effective behaviours which can then form the basis of a training course.

The Repertory Grid technique has its roots in personality research and can often seem rather bewildering on first encounter. However, as a way of getting people to put into words what were formerly just vague ideas, it can prove to be very useful.

At the end of the exercise you should have a list of statements related to the topic of investigation. We could get the interviewee to rank-order the list once the task is complete. This would give us an idea of significance. We could then use the Critical Incident approach and get the interviewee to give an example from the workplace of each of the items on their list. The Grid technique is useful in generating rich descriptions of performance but does require a little practice.

*Card Sorts*

Once we have our list of behavioural descriptions, the final technique we apply is the card sort. Write the individual elements on separate cards and get a set of expert judges – line trainers or managers usually – to sort the cards into piles of associated behaviours. Once the task is complete, get the candidate to explain why groups of cards are related. By this method we can start to get at the conceptual structure being applied. At the end of a card sort exercise you should have a set of related ideas together with some possible labels for behavioural clusters. The card sort is a useful technique for bringing order to apparent chaos as well as useful way of opening up new lines of enquiry.

The process generated a large number of behavioural descriptions that were then arranged into clusters by 'expert witnesses' using a card sort technique. The process yielded 5 major categories of behaviour of which 2 are illustrated above. The first thing to note is that the descriptions of behaviour are much richer than those in the previous 2 schemes. The words used are those of the line pilots and I have tried to describe the behaviour the way it is seen on the line. In a sense, I have tried to look at the 'natural history' of CRM. The way the descriptions have been clustered would probably fail a test of validity and reliability but this, in fact, highlights the problem with soft skills. Behaviour is a mess! It doesn't fall into neat categories, hence the different ways of cutting the cake we have seen so far. The challenge of a CRM assessment scheme is to identify discrete, observable and significant behaviours to be used as valid and reliable measures. In order to meet that challenge, we need to strip out the richness and fuzziness of real-life. Unfortunately, it's real-life that forms the basis of any soft-skills training course. This final set of markers can be used to provide the raw data for subsequent manipulation if we wanted to come up with a more rigorous *measurement* system. In the meantime, it goes a long way towards giving us a foundation for training the skill sets associated with each cluster. We will use this material in the next chapter.

Of course, we have only been talking about pilots so far. We need to ask what behaviours are critical for other key operational groups and we should also ask what behaviours are essential for collaborative working. We applied the techniques described above to some cabin crew and here is the first stage analysis we arrived at:

Cabin Crew Behavioural Markers

A. Approach to Job

Comfortable in Role – Uncomfortable in role
Confident in job – Lacks confidence in own ability
Maintains correct distance – Too close to passengers
Enthusiastic – Unenthusiastic
Cheerful – Gloomy

B. Management Skills

Communicates well – Communicates poorly
Flexible outlook – Rigid outlook
Stays ahead of events – Always behind events
Well organised – Confused
Uses initiative – Lacks initiative
Works hard – Lazy
Remains calm – Easily stressed

Meanwhile, markers are being proposed for Licensed Aircraft Maintenance Engineers (CAA 2003). Here is the element that describes Leadership:

Leadership (Inspiring teams and individuals to better performance)
a. Does not wait to be told what to do but energetically gets on with the job in hand, needing little or no supervision.
b. Actively encourages others to achieve or exceed their objectives, guiding them through challenging situations and difficult problems and publicly applauding their efforts and successes.
c. Motivates others by setting a role model to others through exemplary behaviour and quality of work.
d. Is not afraid to ask for help when needed and accepts advice constructively.
e. Takes personal responsibility for ensuring that tasks are fully completed.

Finally, in this final round up of approaches to markers, here are some generic markers that will apply to all personnel involved in team working. The idea behind this approach is that collaborative teamwork requires individuals to consider the operational needs of others not just now but at some stage in the future. The collaborative behaviour skill set comprises 4 clusters, 2 of which are similar to those found in role-specific analyses: Task Management and Communication. The other 2 clusters reflected the need to take a systems view, what I call 'Keeping the Big Picture', and to recognise the fact that we have a responsibility to support others in achieving their specific operational goals. Here is the 'strategic view' set:

Keeping the Big Picture

Interface between groups
Understand logistical requirements
Consider working needs of others
Understand work processes of others

There is one cluster of behaviours notable by their absence from all the previous discussion. The types of behaviour we are concerned with are the product of learning and, as such, are subject to voluntary control. The behaviours missing from our framework are those we could categorise as self-awareness. By self-awareness, I mean those processes by which we monitor the progress of interactions and selectively chose how to continue the interactive process. By making self-awareness an area of interest, I am recognising that humans are not automatons. I have already said that we exercise agency in that we can control the extent to which we engage with the process of work, comply with procedures, act in a well-mannered fashion and so on. This category of self-awareness throws up some problems for training design. If we consider why people act in an inappropriate fashion we can probably come up with reasons such as misreading the signals, did not know how to react, the performance was sub-optimal, the individual elected not to participate in the interaction. The history of CRM is littered with discussion about those in the pilot community who reject the CRM message: the boomerangs or drongoes. In fact, these are perfect examples of individuals who lack, for whatever reason, the capacity for self-awareness. One aspect of this issue could be that, as I discussed in our exploration of risk, it is likely that the messages of CRM are at odds with the observed reality of these individuals. Their technical skills have never been challenged and their performance has usually enabled them to gain senior positions within their airlines. What, then, do they stand to gain through training in CRM?

When we talk about effective behaviours, we need to support the message with clear guidance on what is effective, why we need to act in a particular way and, finally, what are the consequences of choosing an alternative mode of interaction. We will come back to these issues in later chapters but, for now, I simply want to suggest that existing approaches to CRM assessment, in particular, and training are fragmented and incomplete.

## From Skill Set to Competence

By now you are probably hoping that a simple solution to 'observable behaviours' will fall into place. In fact, 'behaviour' is something of a chimera. I have presented the above lists with little commentary. Although I have not offered complete frameworks, simply examples, common themes can be seen in the different offerings. At the start of this chapter I made passing reference to the idea of competence-based frameworks. The competency movement is rooted in a belief that workplace performance is underpinned by a generic set of competences. For example, numeracy and literacy are fundamental skills required of all working adults. The lack of these basic skills puts individuals at a severe disadvantage in the labour market. Different jobs require different standards of skill. So, accountants need more than the basic level of numeracy, managers in general need to be fairly literate. The competency movement argues that these fundamental skills are transportable across industries and that labour market flexibility will be improved if we can do more to develop basic skills, the argument being that increased basic skills would require less role-specific training for an individual to become competent in a new workplace. This is not the place to debate the rationale behind such a philosophy. Instead, I want to use the idea to develop the broader concept of skilled performance.

The idea of competences is well illustrated by the work of cabin crew. We can divide their work into a set of skills around sustaining safety, a set of skills around service delivery and a set of teamwork skills associated with the collaborative aspects of the job. Within each cluster we can identify role-specific skills, for example how to use a fire extinguisher or how to operate a door, and some broader social skills, such as communication or building effective working relationships. The world of air taxis and corporate jets also reveal the breadth of the competence concept. The relationship between the pilot and the passenger in these types of operation is considerable closer and more intimate than in larger aircraft types. The Captain of the aircraft often has customer-facing responsibilities. The lack of a flight deck door means that pilots are aware that their actions are being observed and this has an implication for the way the crew process is managed.

At this point in the evolution of CRM we have only just started to get to grips with the concept of identifying effective behaviours in a meaningful way. Moving to a competence-based framework is probably premature but I hope this short discussion has shown that we will not have a complete solution to training in aviation until the work of all personnel is defined in such terms. As part of our preparation for course design, we need to explore all possible areas of employment. For example, how do pilot soft skills differ between normal and non-normal situations? What skills are required to deal with situations on the ground before flight?

## Using Markers to Develop Training

This first section of the book attempts to set the scene for Section II, in which we get to grips with developing training. We have covered a lot of ground, some of which might not have been what readers were expecting. So, before we move on, I want to take stock by identifying the problems we need to take into consideration as we continue on to the more practical aspects of our work. As I see it, any meaningful training intervention will need to reconcile a number of conflicting issues.

First, we have the nature of the behaviour we want to foster within our personnel. From the discussion so far it seems that the best way to view worker behaviour is in terms of a set of overlapping clusters moving from specific to generic, from task-related to socially-facilitating. The models offered so far have tended to concentrate on the specifics of completing the assigned workplace role in a safe manner. Historically we have concentrated on a very task-focussed approach to job description. We have learnt, through experience, that this approach is too narrow but we have yet to adequately describe workplace competence in a way that will encourage the design of training that contributes fully to safety and efficiency.

Next, we need to reconcile the role of the organisation. I have said that organisational factors are increasingly being accepted as contributory factors in accidents. The reality, of course, is that organisations provide the context within which workers perform their daily routine. On the one hand, recognising the importance of organisational factors gives us a new target audience for training. The need to consider the impact of management practices offers rich pickings for a new breed of consultants. But, perhaps more important, when we consider the need for a training intervention, we must weigh up that solution against alternative solutions – which could involve changing the organisation.

Finally, we must consider the issue of how we construct risk and safety in our minds. We have seen that statistical models are often at odds with the experienced world. Therefore, our training cannot rest on a set of industry-wide figures in order to reinforce the benefits of a particular behavioural solution. Our training needs to explore the world as held within the individuals in our training cadre and how that world is constructed in association with peers and within our organisation.

In many ways, these are the challenges we ought to be accepting when we sit down to design training rather than simply scrutinising published guidance to ensure compliance. With that in mind, we now need to consider the relationship between a behavioural framework and a soft skills training course. The discussion so far has been based around ways of identifying and describing significant workplace performance. It follows that using the closed-loop training design model the behavioural framework will form the basis of instruction and also the standard for assessment. In effect, we cannot expect personnel to deliver the performance required unless we have trained them first. The testing phase will be dealt with in Section III but, for now, we are interested in the training. In Chapter 1 I referred to the idea of CRM as a lens and as a toolkit. In the same way, the behavioural framework allows us to identify where explanation of behaviour can accelerate the achievement of competence and where the rehearsal of behaviour may be required. If I go back to the competence model, the rehearsed behaviours are the competence and the background explanations are the underpinning knowledge that supports competent performance.

The markers illustrated above are all sub-sets of the full range of behaviours displayed in the workplace. Each cluster represents a set of actions that could be displayed during the accomplishment of workplace tasks. Some of the behaviours are related to the work process while others are involved in fostering good relations and cooperation within the work group. Once we have established our inventory of behaviours we then need a mechanism for assigning a level of significance to each one. Given that we do not have an infinite amount of time and money with which to train our people to proficiency, we need a way of identifying those behaviours that will have the greatest impact on operational performance.

We can fall back on the concepts of risk and efficiency to help us here. We can scan our behavioural inventory and ask to what extent the failure to demonstrate an item will result in an increased risk to the operation. Of course, we have already discussed the insidious, unpredictable nature of risk. So the activity I have just suggested entails a risk! However, if we consider the concept of 'obvious risk' then we can accept that our judgements will be valid. We can also set our judgements against know behavioural failures in accident and incident reports and we may even be able to use company reporting schemes to inform our decisions.

Having applied the 'risk' test, we can go back to our list and ask what cost would be incurred if people do not exhibit the desired behaviour – the 'efficiency' test. If a behaviour cannot be seen to be a barrier against unnecessary risk or preventable cost, then we can assign it to the category of 'optional extra'. I have tried to illustrate how we can use risk and cost to make training decisions in Table 3.2.

Once we have our prioritised list to hand, we can cross-refer against any published regulatory requirements to make sure we have covered the compliance aspect of our work. We will take this process further in the next chapter when we start building our training course.

**Table 3.2 Training Decisions Matrix**

| High | Priority 1 | Priority 1 | **Essential** |
|---|---|---|---|
| Medium | Priority 2 | **Essential** | Priority 2 |
| Low | Ignore | Priority 3 | Priority 2 |
| Risk ⟍ Cost | Low | Medium | High |

**Conclusion**

CRM training needs to be driven by a clear understanding of what constitutes efficient and safe workplace performance. In order to design effective training, we need a complete and accurate competence framework. We also need to understand what factors will prevent people acting in the desired manner in the workplace. Safety and culture are terms we use freely in aviation but I have tied to demonstrate that each individual creates safety and culture for themselves within their workplace. Our training needs to support that process and direct it along the preferred path. The approaches adopted so far in the industry do not meet the standard I have set here but, nonetheless, we will use a sort of competence framework to direct our discussion of training methods in Section II and to inform the discussion of assessment in Section III.

## What to do Next

We have reached something of a watershed in our discussion of building effective training interventions. In the next section we will get to grips with what will happen in a training situation. Before then, we need to assemble some resources. First of all what descriptions of behaviour already exist? Are there job descriptions or terms of reference for any personnel? Where else could you find information? Is there anything in the manuals?

Next, what safety or efficiency data is available? What accident or incident reporting takes place? Is there any form or safety reporting? How about information about delays that could throw light on teamwork deficiencies? What about data on workplace injuries?

Does your airline conduct staff opinion surveys? What is the staff turnover rate like? Do you conduct exit interviews that could provide information about attitudes?

What do management want from this training? Is there consistency of message between what your course preaches and how your airline management behaves (no laughing please!)? Once you have started to gather this data you will be almost ready to move on.

## References

Anon    www.caravanpilots.com

Fegetter, A.J. (1990), 'The Road from Delphi or Hindsight is too late', *Airclues,* January.

CAA, CAP 737

Hodson, R. (2001), *Dignity at Work,* Cambridge University Press, Cambridge.

Hofsteede, G. (1991), *Cultures in Organisations,* McGraw-Hill, London.

Helmreich, R.L. and Merritt, A.C. (1998), *Culture at Work in Aviation and Medicine,* Ashgate, Aldershot.

## Additional Reading

Parker, M. (2000), *Organisational Culture and Identity,* Sage, London.

## Note

1    After I wrote this, the NTSB issued a safety notice covering icing and the Cessna Caravan. Nothing to do with my analysis, I must add!

# Part II
# The Conduct of Training

# Introduction

In the previous 3 chapters I have tried to give some idea of the complexity of the task we are embarking upon. In this section I want to look at the mechanics of solving the problem. In the first chapter in this section we will look at the development of a CRM course. I want to discuss the process of designing and building the various events we may want to incorporate in our classes. In the second chapter I want to look at the delivery of training. Many readers who are already experienced instructors will be familiar with the content of this section. However, I have tried to concentrate on the specific problems most people encounter in the domain of CRM.

I will structure this section around the idea of instructor competencies; the behaviours demonstrated by competent practitioners. I have used the UK CAA 'Guide to Performance Standards for Instructors of Crew Resource Management Training in Commercial Aviation'. There are other, and better, competence frameworks available around the world but the CAA 'Guide' has been around long enough to have been adopted as the default standard within many European countries. I have provided a summary of the standards for ground school instructors at the end of this section (see Table 5.3)

It is important that, as you start to design your CRM course, you keep in mind the ultimate goal of influencing safety within your airline.

# Chapter 4

# Developing Training Activities

## Introduction

Aviation is a small world and there is a strong tradition of self-help and collaborative support between competing airlines. As a result, much training consists of material simply copied from other sources. The very first CRM course to be run in the UK was conducted by a pilot who had attended a 5-day course in the United States and trimmed it to 3 days for the British pilot audience. I need to stress, from the outset, that training material production takes time. As a rule of thumb, you should allow 20 hours of production time for every hour of delivered classroom instruction. As soon as you start to move into more dynamic modes of teaching, then the production ratio increases.

In this chapter I want to explore the process of creating training events from basic principles. We will cover the full range of classroom activity as well as looking at the design of simulator exercises. In so doing, the following competencies from the Guide will be covered:

## Table 4.1 Facilitator Competences

### Plan and Design Training

- A1.1 Identifies training requirements
- A1.2 Identifies design and delivery resources
- A1.4 Incorporates variety of media etc
- A1.5 Involves other people in design
- A2.1 Identifies and selects CRM learning support material
- A2.2 Ensures written and visual support materials are clear, accurate, practical and user-friendly
- A2.3 Ensures activity and exercise materials are practical and realistic
- A2.4 Prepares and presents durable support materials
- A2.5 Promptly identifies and rectifies problems

In developing our course, I want to propose an 8 Step Process. In this chapter we will look at the first 4 steps, which cover the hardcore activities associated with course development. In the next chapter I will look at the final 4 steps, which are largely to do with the supporting activity around the final product. We will also look at the techniques of delivery in the next chapter. The 8 steps, in fact, represent a checklist for course production.

## Table 4.2 The 8 Step Process for Course Production

Step 1 – Clarifying Objectives
Step 2 – Gathering Content
Step 3 – Selecting Activities
    Step 4 – Sequencing Events
    Step 5 – Developing Lesson Plans
    Step 6 – Preparing Visual Aids
    Step 7 – Developing Handouts
    Step 8 – Checking Facilities

## Managing the Process

Before we start, I want to say something about the management of instruction. In any quality management system we need a product specification and training is no different. There are 3 key documents required to support training: a curriculum, a syllabus and a lesson plan. A curriculum is a description of a course. It contains timetables, lesson titles, summaries of content, descriptions of activities, details of any testing regime and so on. A syllabus is a list of training objectives. Given a copy of the syllabus and the curriculum, a new instructor could then prepare a set of lesson plans for the sessions for which they were responsible. A lesson plan is a document used by an instructor to manage a training session. The lesson plan is personal to the instructor and contains whatever detail and guidance is deemed appropriate for the individual concerned. I have worked in many organisations where the lesson plan is the guiding, or often the only, document in existence. Now, of course, it could be possible to generate a single document that achieves all the functions described in one. However, such a tome is likely to be unworkable in class. Also, we need to remember that the curriculum and syllabus are centrally controlled whereas the lesson plan is a working tool. Because each instructor constantly updates and modifies their personal lesson plans, there is a danger that, where more than one instructor covers the same topic, lesson content will start to diverge if the organisation lacks a benchmark for training content.

Using the lesson plan as the control document introduces a further problem. Facilitators bring their own personal experience and style to the learning event. We do not expect all facilitators to act like clones. A centralised lesson plan can be rather stilted in that the facilitator tries to make sense of a lesson written by a third party. An argument for a central lesson plan is that it allows a session to be run at short notice by an unfamiliar instructor in the event of sickness or a forced change of plan. However, this is a contingency and should not be used as an excuse for bad practice.

This chapter assumes that the organisation has done some of the analysis required to generate a syllabus and is now in the process of developing the curriculum. The broad guidelines have been passed to the facilitator who is now going to sit down and prepare their course module and personal lesson plan. To illustrate the process, I want to work through a specific course module. We are going to develop a training element around the topic of Workload Management.

*Step 1 – Clarifying Objectives*

The guidance on topics to be included in a CRM course offered by regulatory authorities, as we discussed in a previous chapter, is minimal. The authorities' argument is that it is for an airline to decide the detail. We have already said that the objectives will have been described in the course syllabus – but let's pretend that our airline hasn't got to that stage yet. What, then, is an objective? An objective is a statement of the learning to be accomplished by the trainee at the end of the session. Objectives are usually written as performance statements. So, we might write:

> At the end of training the student will be able to…

We complete the statement with some description of the action required of the student to demonstrate that they have, indeed, learnt something during class. Writing training objectives is as much an art as a science. It is also time consuming. However, it does bring discipline to the training process. If we look at the published guidance, what are we supposed to cover in a session on Workload Management? The FAA offers the following (FAA 2004):

> **(3) Workload Management and Situation Awareness.** Stressing the importance of maintaining awareness of the operational environment and anticipating contingencies. Instruction may address practices (e.g., vigilance, planning and time management, prioritizing tasks, and avoiding distractions) that result in higher levels of situation awareness. The following operational practices may be included:
>> **(a) Preparation/Planning/Vigilance.** Issues include methods to improve monitoring and accomplishing required tasks, asking for and responding to new information, and preparing in advance for required activities.
>> **(b) Workload Distribution/Distraction Avoidance.** Issues involve proper allocation of tasks to individuals, avoidance of work overloads in self and in others, prioritization of tasks during periods of high workload, and preventing nonessential factors from distracting attention from adherence to SOPs, particularly those relating to critical tasks.

Meanwhile, the relevant JAA publication simply says:

> Workload Management

As they stand, neither document offers us enough detail with which to build our course and, so, we need to elaborate on these themes. The first place to start would be our behavioural markers as these represent how workload management would be manifested in the workplace. On page 48 I described the Task Management cluster of behaviours developed through my work with line pilots. If we analyse this description, we can pick out some component parts of the behavioural cluster. For example:

'Consistent and appropriate use of checklists and procedures'
'Monitoring the use of checklists during busy periods and the positive verification that tasks have been completed'
'Maintaining an even tempo of work (no unnecessary haste or urgency)'

'Making effective use of time'
'Anticipation and thinking ahead'
'Being aware of time available/remaining'

'Maintaining the bigger picture of things happening around the aircraft'
'Actively developing mental pictures of what to expect during the next stage of flight (e.g. through verbalisation of expected landmarks, events, system changes etc)'

I have arranged these statements into 3 sets that, to me, form coherent groups. The first points at the control of the work process. The second group is all about time management and the third is, perhaps, what we could call situational awareness. It is interesting to set these clusters alongside the earlier discussion of the Little Rock event.

If we recall the lens/toolkit dichotomy, we now need to consider what people need to KNOW in order to deliver the performance and what people actually need to be able to DO. Looking at the first group of statements we can do a little brainstorming around what the goals of training might be in association with these observed behaviours. We might generate a list like this:

Need to know how checklists and procedures are written.
Why do we have checklists?
Strengths and weaknesses of checklists.
When can we be flexible in applying checklists and procedures?
Where are the risks in using checklists?
How do we manage checklist interruption?
What constitutes poor checklist/procedure application?
What traps do we fall into with checklists or procedures?
How do we solve problems using checklists and procedures?
Accept the need for checklists.
Interpret procedures correctly.

Writing training objectives is tedious. However, the process is, on the one hand, an act of discipline and, on the other hand, a quality control measure. We need to take the output from our brainstorming exercise, together with the regulatory guidance, and identify those items we want to take further and establish as outcomes of learning. A training objective, ideally, contains a statement of the action a student should be capable of demonstrating at the end of the training intervention. Historically, we have tended to allocate outcomes to 3 broad categories: skills, knowledge and attitudes. A skill is some process that can be observed. Knowledge acquisition is accessible through questioning or through formal testing. Attitudes reflect a disposition to act in the desired way and can be accessed through questionnaires or by observing work place performance. If we look again at the behavioural elements extracted from the marker scheme, some of these points relate to action – 'interpret procedures correctly' – while others are to do with attitudes – 'accept the need for checklists'. Other points relate to exercising judgement – 'when can we be flexible...' – while others require an understanding of the basic principles of checklist design. As they stand, these statements do not tell us what the desired outcome of training might be. We now need to develop the second part of the training objective. At the end of training the student will be able to...what?

Because we are interested in verifying that learning takes place, the heart of a training objective is a statement of performance, usually couched in terms that represent observable outcomes. To show what I mean, let's assume that, in order to train all students to a minimum standard in procedural activity I may want to develop the concepts implied by the statement:

What traps do we fall into with checklists or procedures?

Through my initial research I may have identified a number of categories of checklist malpractice, some of which may be the result of poor checklist or procedure design. I want to make sure my students are forewarned of the dangers associated with the use of checklists and so I can write a training objective that says something like:

At the end of training students will be able to describe the $x$ common failures in checklist application.

But I want students to be able to do more than just tell me what could be wrong with using checklists. So, I might write an additional objective that says something like:

At the end of training students will be able to describe $y$ techniques for ensuring safe application of checklists.

We are still in the realm of 'telling' rather than 'doing'. The first objective reflects underpinning knowledge and, while the second objective is closer to the realm of activity, being able to tell me what they would do is still no substitute for students actually using checklist. I might, therefore, complete this block with an objective that says something like:

At the end of training students will be able to demonstrate safe application of checklists and procedures.

None of the proposed objectives target student attitudes towards the use of checklists. After one accident in the USA, investigators flew with other company pilots and reported that pilots seemed uncomfortable using checklists. I have even heard the view stated that 'checklists are for wimps'. The attitudes people bring to regulated behaviour are important and so we might want to recognise that fact in our training – and you only need to think back to our discussion of drivers and speed limits to understand what I mean. Therefore, we may want to add an objective such as:

> At the end of training students will recognise the need for appropriate checklist application.

Note that, so far, we have made no decision about how we propose to deliver the training. Nor have we made any allowance for the skills students may bring to the class. Our goal is simply to describe what we want to achieve during training. The student entry level – the competence they bring with them to class – will be taken into consideration once we get closer to making decisions about training methods.

If you look back at my brainstormed list, you can probably see that certain statements stand alone as possible training outcomes while other statements seem to have a supporting role or possibly reflect the content of a session dealing with the specific learning outcome. As the developer, you need to make some decision about where to draw the line in terms of defining objectives. Each objective selected should, in some way, contribute to the development of our personnel. As a test, an objective that we can look at and say 'so what' is probably not a good objective. I have included some specimen objectives for the key CRM topics at the end of this chapter (pages 109-110). They are not offered as exemplars, simply short cuts!

Before we move on, it might be beneficial to bring some closure to this discussion of objectives by establishing our goals for the workload management module. In Table 4.3 I have arranged the ideas developed in our discussion so far. It is not a compete framework, nor is it offered as a perfect solution. You will discover for yourself that objectives only ever represent a way station on the route to perfect statements of learning, the destination constantly retreating over the horizon.

## Step 2 – Gathering Content

Training objectives identify the outcomes of training in terms of performance. They also identify the areas of course content we need to develop. Gathering content for the lesson requires diligence and, sometimes, ingenuity. With the internet becoming an increasingly useful resource, in some ways we can have too much information available. Selection becomes almost as difficult as finding the information in the first place. As part of your task as course designers, you need to take the official guidelines and see how they apply to your airline. We will be dealing with lesson planning as an activity in the next chapter but, for now, I want to introduce a planning document which bridges the gap between identifying our training goals and developing lesson plans.

## Table 4.3 Workload Management

At the end of training students will be able to

1. Describe the purpose of checklists and procedures.
   Why do we have checklists?
   Strengths and weaknesses of checklists.
   Need to know how checklists and procedures are written.
   Accept the need for checklists.

2. Describe the $x$ common failures in checklist application.
   Where are the risks in using checklists?
   What traps do we fall into with checklists or procedures?
   Interpret procedures correctly.

3. Describe $y$ techniques for ensuring safe application of checklists.
   What constitutes poor checklist/procedure application?

4. Demonstrate safe application of checklists and procedures.
   Behavioural outcomes:
   Consistent use of checklists and procedures.
   Appropriate use of checklists and procedures.
   Monitoring the use of checklists during busy periods.
   Give positive verification that tasks have been completed.

5. Recognise the need for appropriate checklist application.

6. Describe the techniques available to manage workload.
   Proper allocation of tasks to individuals.
   Avoidance of work overloads in self and in others.
   Prioritization of tasks during periods of high workload.
   How do we solve problems using checklists and procedures?

7. Demonstrate workload management.
   Behavioural Outcomes.
   Making effective use of time.
   Being aware of time available/remaining.
   Maintaining an even tempo of work (no unnecessary haste or urgency).
   Anticipation and thinking ahead.

Supporting Topic – Situation Awareness.
'Maintaining the bigger picture of things happening around the aircraft'.
'Actively developing mental pictures of what to expect during the next stage of flight (e.g. through verbalisation of expected landmarks, events, system changes etc)'.

A copy our proposed lesson Planning Sheet is at the end of this chapter (page 111) but, basically, we need to answer 3 questions. Having extracted the topics from the syllabus, we need to ask:

- What does it mean for us?
- Do I have examples from my airline?
- Do I have examples from other airlines?

As we gather the answers to these questions, we are elaborating our understanding of what performance we expect of trained crews and, therefore, we can deal with the 'what' and 'how deep' problems identified earlier.

One method of content generation is the technique of brainstorming. We might have done some sort of brainstorming in order to develop our set of training objectives. This requires us to sit down with a group of colleagues and generate lists of ideas, words and concepts all linked to the topic. We can include personal examples where appropriate. Ideas are written on a whiteboard or flip-chart until the group has exhausted its pool of knowledge. Ideas can then be clustered around common themes. The Safety and Quality managers are sources of useful information, as are accident and incident databases. Some useful websites are listed at the end of this chapter.

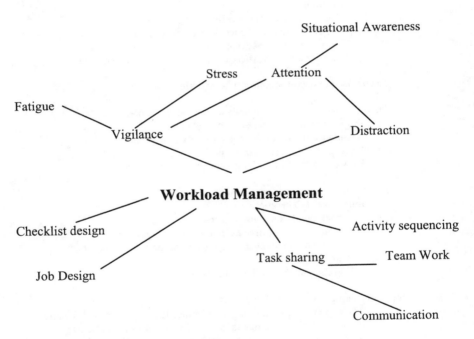

**Figure 4.1 Topic Concept Map**

At this stage we are simply collecting resources. How we use material to construct a lesson will be dealt with later. In the case of our demonstration module, the information available on workload management within aviation is not extensive and so we may need to look at other workplace settings. What we will quickly discover is that there are many ways to divide up the target knowledge-base. The boundaries between some topics are very fuzzy and, to a degree, we are providing overlapping coverage of content rather than discrete parcels of knowledge. We can use mind-mapping techniques to show relationships between ideas and, having sketched out relationships, make decisions about where to draw boundaries. In Figure 4.1 I have tried to illustrate this problem in relation to our example topic.

*Step 3 – Selection of Activities*

By now we should have a good idea of the objectives we intend to cover in training and we have started to collate suitable source materials. We can now turn to developing the individual training activities. There are a variety of vehicles available to us for the delivery of instruction, each having its own strengths and weaknesses. I have listed the most likely training events we will want to use in Table 4.4 and have attempted to summarize the key strengths and weaknesses of each one.

**Table 4.4 Strengths and Weaknesses of Classroom Activities**

| Activity | Strengths | Weaknesses |
|---|---|---|
| Lecture | High information transfer. Works with large groups. | No interaction. Low retention. |
| Lesson | Allows interaction. Easily controlled by facilitator. | Moderate information transfer. |
| Case Study | High relevance and interest. | Selection and administration. |
| Questionnaire | Provides individual feedback. | Slows pace of class. |
| Video | High interest. | Requires technology. Fits lesson objectives? |
| Role-play | Good for behaviour modification. | Time consuming. Some rejection by audience. |
| Practical Exercise | Establishes common standard. Interactive. | Time consuming. |
| LOFT | High face validity. | High production costs. Requires extensive stage management. |

Selecting a training event – or medium – is an exercise in compromise. On one hand we would hope to match the stimulus presented in class to the output required in the real world. The closer we can replicate actual skills, the better the training should transfer to the real world. At the same time, we need to consider the cost/benefits associated with each activity. In Table 4.2 I have rank-ordered the activities very loosely according to their cost of production (assuming you buy a video and not produce one from scratch – but see Table 4.5). Part of the compromise is selecting a method of instruction that delivers the best return in terms of training transfer for a given cost. What we are aiming for is a mix of events that contribute to the development of understanding on the part of course members or that allow delegates to apply or extend skills. At the same time, we need to do this in a relatively short space of time. The task of the course designer is to select the optimum activity for the objectives to be met.

If we return to our demonstration module, Workload Management, we need to examine the objectives in terms of the intended output performance. We can see that we want to convey some information about the process as well as allowing the class to rehearse workload management techniques. We have also said something about wanting to reinforce attitudes towards checklists and procedures. The main types of training activity listed in Table 4.4 are evaluated in terms of training outcomes in Table 4.5. We need to remember that the categories are not mutually exclusive. For example, a LOFT scenario in a simulator could be considered a form of role-play. Describing a drill in a classroom is a form of mental rehearsal and could, at a stretch, be considered a simulation. The categories, then, are fairly loose labels but, nonetheless, recognisable to readers as things they have probably endured during their own training.

**Table 4.5 Training Inputs and Outputs**

| Input | Output |
|-------|--------|
| Lecture | Head full of facts |
| Lesson | Partly-digested facts |
| Case Study | Facts and behaviours in context |
| Questionnaire | Self reflection |
| Video | Dynamic representation |
| Role-play | Emotional component and behaviours |
| Exercise | Skill rehearsal |
| LOFT | Skills and procedures |

From Table 4.3 we can extract 3 action verbs associated with the anticipated output from training: describe, demonstrate and recognise. The first 2 verbs lead us towards situations where students can discuss topics and rehearse skills. The third verb, recognise, calls for a different type of activity. This objective targets student attitudes and here we enter hazardous territory. Historically, attitudinal objectives have been consigned to some sort of wish list in that it has been accepted that reliable evidence of attitude change cannot be garnered over a period as short term as a typical training course. Whereas skills and knowledge can be directly targeted in training, attitudes have to be stalked. Attitude change will often result from increasing an individual's understanding of the object of the specific attitude and so we may bring about change as a spin-off from other activities.

As part of our media selection process I said earlier that we need a way of mapping these desired outcomes onto the available training methods. In Table 4.6 I have tried to do just that. Of course, it is difficult to establish a perfect fit between training methods and learning outcomes. There is a host of intervening variables, such as entry standard, personal motivation to learning, learning style, which will influence the extent to which an individual student will benefit from our efforts. I discuss these in more detail elsewhere (MacLeod). For planning purposes, though, the Table, which also tries to show relative strengths of each medium, should work.

**Table 4.6 Media Selection**

| Medium | 'Describe' Knowledge? | 'Demonstrate' Skill? | 'Recognise' Attitude? |
|---|---|---|---|
| Lecture | √√√ | | |
| Lesson | √√√ | | |
| Case Study | √√ | | √ |
| Questionnaire | √√ | | √√ |
| Video | | | √ |
| Role-play | | √√√ | √ |
| Exercise | √ | √√√ | √√ |
| LOFT | √ | √√√ | √√ |

Armed with information about effectiveness in relation to learning outcome, cost of production and time available for training, we can begin to select appropriate learning media. I have been allocated half a day for my Workload Management module. I cannot find a questionnaire that identifies respondent's attitudes to rule-following behaviour, nor do I have the funds for a video. I have decided that I want to use a case study, a lesson and a practical activity of some sort. I'd considered a role play but decided against it because I want to focus more on the skills of managing tasks as opposed to the social aspects of collaboration during task management. A LOFT exercise, albeit in a generic classroom-based format, might work but I don't want the class to get too hung up on procedures. Furthermore, the course may be delivered to newly hired students who do not yet understand our operation.

On balance, I have decided that some sort of practical classroom activity would be best. I have made a note to co-ordinate with our simulator instructors to see how we can reinforce the learning during the simulation phase. From Table 4.7 I see that this module is likely to take me at least a week, and probably closer to 2 weeks, to construct. I inform the crew schedulers of my unavailability, close the office door and set to work. For the rest of this chapter I want to explore the processes involved in developing each of the key practical activities on our list of training media but, before that, I want to finish this first phase of training design by looking at stringing individual events together into a coherent course plan.

**Table 4.7 Media Production Times**

| Input | Production ration per delivered hour |
|---|---|
| Lecture | 5-15 |
| Lesson | 5-20 (up to 50 if preparing materials for others to use) |
| Case Study | 20-30 |
| Questionnaire | 10 – 200 depending on level of validity required |
| Video | 5 if sourced, 400 if commissioned |
| Role-play | 15-30 |
| Exercise | 10-40 depending on complexity |
| LOFT | 20-50 |

*Step 4 – Sequencing of Events*

Having identified the content of our session and the activities we want to use to support learning, the next stage is to arrange the training events in a sensible order. At the macro level we are interested in sequencing within the training day and at the micro level we want to see how things fit within a single training session.

*Macrostructure*

At the level of the whole course we can see at least 2 factors that influence sequencing decisions. The first is the flow of information and the second is the nature of the individual training events. If we look at the flow of information first, it is sometimes possible to identify a coherent structure that can be mapped onto the various topics to be covered. For example, most initial CRM classes describe a path that includes aspects of the individual, group processes and organisational processes. At the level of the individual we can include topics such as stress, fatigue, human information processing, personality and attitude. Group processes include task management, decision-making, communication and leadership. Organisational topics would include discussion of safety culture and management.

I initially ran courses starting with the individual and passing through the group processes and ending with the airline as an organisation. The course worked well-enough but, for reasons that remain obscure to this day, after reversing the sequence so that we started with the big-picture, systems-view I found that the course was much better received.

When designing recurrent courses that focus on just a few topics it can sometimes be harder to perceive a coherent structure for what can be a disparate group of topics. In these cases it is often simpler to let the activities drive the structure.

The activity-driven approach simply accepts the fact that the planned classroom activity, be it a video, case study or exercise, has to be completed before anything else can happen. So, we undertake the task, debrief the key points and use the debriefing to set the agenda for the subsequent session. For example, the Desert Crash exercise (see p78) is primarily about decision-making. However, groups can only complete the task by communicating with one another. A logical flow, then, would be to do the exercise and lead into a session on decision-making followed by a session on communication.

Under the heading of macrostructure we should also consider the needs of the student in terms of variety of stimulus. We said earlier that we can use the structural elements of the course to vary the tempo of the course, to force the class to engage with the material in an active sense, to get people on their feet and moving around. These aspects of course development are all important if we want to achieve results.

With all this in mind, I have decided on a rough game plan for my Workload Management module:

> Introduction – a short element to get the session going.
> > Case Study – We will look at an appropriate example as a way of scoping the topic and establishing the key issues.
> > Lesson – At this point I will get the class to discuss the problem and introduce some of the theoretical elements.
> > Exercise – We will finish by allowing the class to rehearse some of the skills we have identified in the previous 2 events.

*Microstructure*

Having arranged our topics and activities in some sort of order we now come have to look at arranging material within our individual sessions. A number of rules of thumb have developed over the years to guide us. For example, moving from the concrete to the abstract; from the known to the unknown, from simple to complex; from the particular to the general. We will discuss microstructure in more detail in the next chapter. Now, we will return to the mechanics of media production.

**Creating a Case Study**

Case studies are fundamental to CRM training. Social scientists have woken up to the extent to which myths and tales are used to communicate within organisations, but 'hangar talk' has been an integral part of aviation training since, well, the dawn of aviation. In fact, case studies are the only training event to be specifically included in the regulatory requirements. Despite the fact that the Regulators are confusing content with method, first the UK CAA and then the JAA has required case studies to be part of CRM training, although they both fail to tell you what you are supposed to do with them.

In fairness to the drafters of the requirement, I suppose they were simply reflecting the historical trend. Case studies allow us to put theory into practice – well, almost. We need a way to ground the various ideas included in the CRM domain. We could always ask the class members if any of them have had an experience that relates to the session subject matter but this approach is unreliable in that we cannot guarantee to have an 'eye-witness', so to speak, on every course and some delegates may be reluctant to relive a bad experience. Having spoken to crews who have been through interesting experiences, I have been struck by the extent to which the crew were unable to agree on what actually happened and the extent to which they could not remember what happened. Memory is fallible; memory under stressful conditions can be even more fallible. A case study is really a surrogate experience. Through looking at what others did, we can put theory into context.

I have deliberately included case studies and videos in the same category as they are both representations of an event but presented using different media: broadly speaking, print versus moving pictures. Each has its own strengths and weaknesses and these will be covered in the next chapter. The important point is that, as part of our course, we have decided to examine, in some detail, a specific event. It follows, then, that we need access to information about that event. With the growth of the internet, and with ever more national accident investigation authorities developing web sites, the English-speaking aviation community is gaining unparalleled access to accident and incident reports. So, our first requirement for developing a case study would seem to be easily met.

Unfortunately, getting access to suitable material is only part of the problem. For reasons discussed in Chapter 2, reports can be incomplete or biased. Some accident investigation authorities do try to simply list the pertinent information with no discussion. The problem here is that, without specialist knowledge of the aircraft type or the nature of the operation, the meaning of the stark data can be lost. Very rarely do we get to hear the voices of the participants in the event. Therefore, motive – the reason why people did the things reported – is hidden. Even the best-written reports need to be treated with caution.

*Selecting a Suitable Incident Report*

What makes a suitable report for use as a case study? First, it helps if the aircraft type involved is one operated by your own airline. There seems to be a more immediate identification with the scenario if the class is already familiar with what is going on through a shared experience. Second, a good case study involves an inter-play between technical and social skills. The vast majority of incident reports are simply statements of technical failure. Very rarely do we hear how the crew dealt with that failure unless, of course, they made the situation worse. Third, a good case study in a CRM context is one where the whole crew has been involved in some way. Again, in most reports, we only hear of the actions on the flight deck. And by the whole crew, we should remember that ATC, maintenance and management all play a part in events.

I tried to draw up a framework for analysing accident and incident reports in terms of the 'ingredients' in the narrative – rather like literary analysis. I found 4 main ingredients in a good 'story':

- Precursors
- Trigger Events
- Responses
- Aggravating Actions

The precursors are really latent conditions. Latent conditions are elements of the system lying dormant, waiting for the right set of circumstances before coming into play. For example, in an on-ground evacuation, a member of cabin crew had difficulty opening the Galley Service Door on a Fokker F28. It transpired that cabin crew emergency training rarely involved operating this door which, although a nominated emergency exit, was sufficiently different to the main passenger door to cause problems on the day. The value of having precursors within the scenario is that it reinforces the unpredictable nature of risk and illustrates the need for thorough examination of a problem, time permitting, before deciding on action.

Trigger events are those events that cause the subsequent action. Trigger events can be either human actions or technical failures. Sometimes the trigger event is obscure. For example, a Beech 1900 experienced an engine fire on landing. The trigger event was the crew turning on the landing lights. The latent condition was a failure by the maintenance crews to properly secure the landing lamp wiring to the aircraft structure, in this case the fuel pipe. The crew was presented with 2 indications, fuel low pressure and an electrical failure, neither of which could possibly have been attributed to the trigger event. The trigger event is typically the point of entry in to the scenario.

The Responses, naturally enough, are what the players in the scenario do to resolve the situation. Normally, of course, everyone does exactly what is required unless they particularly want to appear in a case study!

Finally, Aggravating actions are those actions taken by players in the scenario that make the situation worse. Some aggravating actions are failures to execute drills properly. Other aggravators may be innocent actions taken because of the obscure nature of the trigger event. The use of the term aggravator is not intended to reflect on the crew in any way, which brings up an important point. We are not using a case study in order to identify culpability. Case studies fall into the 'lens' category of training. By examining events in some detail we are seeking to understand the 'how' and 'why' of the situation. Case studies offer little opportunity to develop skills. They are not 'toolkit' devices.

*Developing the Case Study*

Once you have found a suitable report, the first thing to do is make sure you fully understand what has happened. The intricacies of procedures and systems functioning may not form part of the subsequent exercise but they could prove to be part of the context needed to understand crew actions. As you read the account, look for signs of bias on the part of the report writer. The biggest failing of accident and incident reports is the fact that the drafters easily fall victim to hindsight bias. In hindsight, it is easy to see what went wrong and, therefore, what the crew should have done differently. You find the same thing happening in CRM classes. It is important to try to understand what the crew was trying to accomplish, what information they had to hand, what options were open to them at the time. Although some of these questions may form the basis of the analysis in class, they reflect the level of understanding necessary on the part of the training developer on order to produce an effective case study.

At this point we need to start thinking about shaping the material. You could, of course, simply hand out the report and let the class set to work. The problem with this approach is that many of the best incidents, and certainly most accidents, are such complex events that there is a problem of too much information. Sifting through the event takes time. We need to edit. Shaping the case study requires us to map the content onto the objective we are trying to meet. Many case studies contain evidence from across the spectrum of CRM topics. We need to decide upon a generalist or a focussed approach. The generalist approach involves presenting a broad description of the event and letting the class draw out the CRM aspects. This approach can be used as a front-end to quickly review the course content. We can also use a generalist case study to support several sessions, returning to it at intervals to focus on the aspects of the case study that are relevant to the particular lesson. The focussed approach involves stripping the case study of all unnecessary detail other than enough contextual data to understand events and the specific CRM behaviours we want to illustrate.

Having decided on how you are going to frame the case study, the next step is to redraft the content. Often reports are written in a formal style, many have been translated from a foreign language. This is really an exercise in effective communication: taking a technical report intended for a different audience and rewriting to suit your audience. In so doing, you need to make sure that you do not miss important information or distort the facts. Some reports can be quite complex and it can be quite easy to cause confusion. You may need to reorder the material in order to make a more coherent narrative. You should also keep in mind the need for any supplementary material or illustrations that might help comprehension.

Finally, I need to sound a note of caution. In aviation we think of fatal accidents as organisational wake-up calls. Fatal accidents represent seismic events in terms of organisational learning and very few such accidents fail to reveal catalogues of shortcomings within the affected airline. As such, it can be easy for the lessons from the subsequent report to be rejected simply because of the uniqueness of the event. I prefer to use more everyday events. Preferably, events when every one was able to walk away afterwards. By using smaller-scale incidents, you are more likely to strike a note of recognition with your audience.

The mechanics of using case studies in class will be dealt with in the next chapter. My goal here was to describe the work involved in developing a useful case study in the first place. I have said nothing about videos. Few airlines are in the position to commission a training video and so are dependent upon material already in existence. As a result, it is rare for a video to match your training objectives and so they are best thought of as being in the generalist tradition. If you use a video, it is still important to go back to the original sources, the investigation report, as it is not unknown for videos to give highly slanted version of events.

## Role-plays

Even after over 20 years in existence, CRM is still seen as threatening by some members of the aviation community. So much so, in fact, that early guidance from the UK CAA stressed that CRM was not intended to change a pilot's personality. It is probably a moot point as to what exactly is being changed but the fact remains, however, that we are trying to modify behaviour, a goal reflected in my toolkit analogy. One of the most direct methods of behaviour change available to us as course developers is the role-play.

Role-plays can be powerful learning opportunities. They provide a safe environment within which participants can try out different ways of behaving, explore the emotional component of interactions and evaluate the affect of attitudes on behaviour. The problem with role-plays is that they can be the one of the last refuges for ham actors!

A role-play has 3 distinct phases. First, we need to establish the situation. This involves explaining the purpose of the activity, outlining what is required of participants, explaining any rules of the exercise and setting any bounds on behaviour (for example, no kicking, biting or scratching). Next, we run the role-play itself and, finally, we bring everyone back to reality to discuss the experience and to consolidate the learning that, hopefully, has taken place. Because the participants are able to act out different ways of behaving and are also forced to confront some of the emotional issues associated with behaviour, this method of training can be very powerful. For the same reasons, though, some people reject role-play. They do not like having to perform in front of others and feel that the artificial nature of simply 'acting' reduces the value of the experience. These problems can be overcome by following the 3-stage structure outlined earlier, by ensuring that the briefing properly sets rules and objectives and, during the debriefing, take some time to dissipate any energy remaining in the class from the role-plays themselves.

Designing the role-play requires us to, first, establish what behaviours we want to explore. We then develop an appropriate scenario that includes some trigger events that will allow the participants to rehearse the behaviour under review. The skills of communication, conflict resolution and, indeed, just about anything to do with working in a group can all be the subject of role-play training. To explore the design process further I want to give an example that looks at the concept of assertiveness.

Imagine that we are presenting a session on Assertive Behaviour. We have dealt with the 3 states of assertive, aggressive and non-assertive, offering definitions of each together with examples of typical behaviour. We could then ask the class to share their own experiences of encountering such behaviour among colleagues. We can also let them act out the styles to experience for themselves, first hand, what we have been talking about.

We brief the class that they are now going to have the opportunity to deal with the different types of behaviour. As part of our planning we will have decided if we want a 1-sided or 2-sided interaction. In the former, we can present the situation to an individual and ask them to give their response to the whole class or, in the latter case, we can set up both sides of the interaction and let them work through the event. For now, we will try a 1-sided approach. Here is the scenario, which is based on a real event:

You are part of the crew of an aircraft. You have had a message from the gate to say that several of the passengers have tried to board with more than the allowance of cabin baggage. In fact, the gate agent has collected enough bags to fill 3 hand trolleys. She has arranged, with the ramp agent, to bring the bags out from the gate and will load them into the forward hold.

Things are already quite tight with respect to meeting your slot time. Most of the passengers have now boarded. The First Officer has seen the gate agent pushing the first trolley out to the aircraft and she is now loading the bags herself into the hold. There are 2 more trolleys to be moved and unloaded. You are just about to go down and help the gate agent when you see 2 male cabin attendants and a dead-heading pilot sitting in the almost empty Business Class. You ask them to lend a hand to get the excess hand baggage stowed so that you can get away on time. One of the guys says that it is not his job to load bags.

First of all, what do you think of the scenario? Is it plausible? Are there any gaps or possible causes of misunderstanding on the part of the role-play actor? Of course, if your airline doesn't have a Business Class cabin than we have a problem of validity. If the method of working doesn't ring true – for example, what's the chance of getting to sit down and read a paper during boarding in your airline? – then, again, the scenario is weakened. Think about the role-player. Who are they supposed to be within the event? We need to test each scenario to make sure the class participants will not spend more time critiquing your design efforts than they do in attempting the exercise. Once we are satisfied, then we can proceed. Having presented the scenario, the actual role-play involves getting the participant to verbalise a response. So we might say to one class member:

How would an aggressive person respond to that guy?

We need to make sure that the participant doesn't attempt to describe how someone might respond, we want them to use the words. We can repeat the process with other class members working through the alternative non-assertive and assertive modes of behaviour. The various responses now provide the raw material for our analysis. We can consider the difference between the styles of communication event and we can explore their relative effectiveness. We can also look at the underlying emotions. So we could then ask:

What would you think about the guy who was not prepared to help?
How would you feel if someone spoke to you that way?
How would you respond to such a person?
Why would you act that way?

We can turn the role-play into a 2-sided event by having a class member play our uncooperative newspaper reader and to let the conversation work through to a conclusion.

We can also run role-plays as group events. In the following example, we can divide the class into small groups. Tell the groups you are about to set the scene and you will be asking them to discuss possible courses of action as the scenario unfolds. Throughout the scenario, refuse any requests for extra support, delays, flight cancellations etc. The flight must depart and the crew must deal with things themselves. Start the exercise by reading out the following background information.

> You reported for duty at 0750 for an 0850 departure and have just landed at a Caribbean airport after a 3.5 hour sector. The outbound flight was without incident other than a disagreement between the Captain and the Senior Cabin Crew Member (SCCM). The Captain had not conducted a crew briefing and during the trip out to the aircraft he had not bothered to introduce himself to anyone on the crew bus. The SCCM had gone up to the flight deck to introduce herself, only to have the Captain comment 'So you are the one who stays in their room for the whole night-stop'. The SCCM replied 'I didn't know I was the topic of conversation' and stormed back to the forward galley to tell the other cabin crew of the comment.
>
> On arrival, you are told that the company wants your aircraft to operate a sector cancelled yesterday due to mechanical problems. Yesterday's crew, who were forced to night-stop, will take your aircraft. You will take over the broken aircraft. However, spares and a mechanic are already inbound and the broken aircraft should be easily fixed although not in time for your planned return, which is at 1400 local.
>
> The Pilots have gone to Dispatch and the cabin crew are sitting in the terminal building. The mechanic has arrived and is working on the aircraft.

Once you have set the scene, tell the class that you will be providing further information in a moment. At each stage, you want the groups to discuss the information and to identify possible options.

The time is now 1430. Tell the class that the mechanic has come back and said that the aircraft can now be released to service. There is some paperwork to complete but that should only take another 30 minutes. Now tell the class that:

> A ground handling agent approaches one of the cabin crew and says 'You realise that, if you do not get away soon, we will not be able to cater the aircraft as the kitchens are about to close. There's no water aboard, drinking or in the toilets. If someone doesn't make a decision we'll have no way of fixing things'. The conversation was overheard by a group of passengers who now approach the cabin crew.

Once the teams have absorbed the information, ask them, in their groups, to identify possible courses of action at this point. Get the teams to present their options and their preferred choice. We have now created the entry point for a series of possible role-play interactions between cabin and pilots and between cabin and passengers. However, I need to sound a note of warning. A scenario like the one we have just developed is fairly open-ended in terms of where it might go and the specific behaviours demonstrated. Without trying too hard, we can see that we could be looking at conflict resolution, communication, interaction styles and even leadership. Role-plays need to be carefully designed to match the behaviour under discussion if you want to avoid a general free-for-all developing. You may also have noticed that the initial scene-setting was gender-specific; male pilot and female SCCM. Gender roles are a legitimate topic for discussion in terms of workplace interaction and you may chose to manipulate this aspect of the scenario as part of your planning. In this case, the genders are those of the players in the real-life event I have used as the basis of the role-play.

I said at the start of this discussion of role-plays that this particular method can be used to explore emotions, attitudes and their impact on the subsequent behavioural output which, generally speaking, will be some form of verbal interaction. Role-plays can also act as vehicles for skill development. In the first example we gave – the less than helpful cabin crewmember – the simple act of rehearsing different ways of communicating is, in effect, helping to develop skills. In the next section we will look at a different technique for targeting skills.

## Classroom Practical Exercises

The use of group exercises in leadership and teamwork development has a long and venerable history. Generations of military officers and middle managers have carried heavy weights for miles across rugged countryside or spent hours dangling off cliffs at the end of a rope. Although subject to periodic changes in fashion, practical activities can be powerful training tools and I know of one airline that successfully included orienteering within its 3-day Initial CRM course. Most courses, however, have to cut their cloth to suit their circumstances. That said there are still many ways to deliver practical training inside the classroom. Group exercises can serve 2 distinct purposes and fit neatly into the lens/toolkit paradigm. We can use an exercise to allow the class to demonstrate behaviour – toolkit – and we can also use an exercise to force the class to contemplate some aspect of the syllabus – the lens approach. The main difference between an exercise and a role-play is that, in the latter, participants have to simulate actual behaviour – they have to pretend to some degree. In an exercise, what we are seeing is, in the main, actual behaviour. I want to work through some of these issues by looking at a classroom exercise in some detail. The example I will be using has been run for airlines around the world, and not always successfully. The full exercise is available at the end of the chapter and can be used by readers in their own courses.

Before looking at the example in detail, though, I just want to deconstruct an exercise in much the same way we analysed case studies. I mentioned outdoor management training earlier. I once worked as a leadership instructor and have spent many a happy day blundering through fog and rain behind groups of potential leaders as they struggle to hold a map the correct way up. Every exercise we did had 2 basic ingredients: a briefing and a task. The task can be further subdivided into a problem solving phase, a transit to a location and an implementation phase where the original solution is put into effect.

We will consider the briefing first. The briefing phase serves to convey enough information to allow the team to understand what is required but should also presents a challenge in its own right. The briefing represents the point at which the team starts to work together. By building in either complexity or ambiguity, we can put pressure on the team from the outset. In the first exercise we will be looking at, we present the information in a written format but on 2 sheets of paper. This requires the group to recognise that the complete information is spread across 2 sources and needs to be integrated. The exercise can start without this taking place but the team will not succeed.

The other common feature of all practical exercises is the task. Now, that probably sounds like a statement of the blindingly obvious but we need an activity that will keep the team occupied for a period of time. The task itself is not the objective; it is simply the vehicle within which the team demonstrates behaviour. We can manipulate the task in order to vary the complexity, time available, number of people involved and so on. Aspects of the task put a burden on the team. If we go back to outdoor training, we could increase the stress levels by varying the distance between the start and the point of task execution; by restricting the time available; and by varying the quantity of equipment that needed to be carried. The task, then, is the engine that drives the exercise along and which generates the behaviour we want to examine. The task can vary in its level of fidelity. For example, and notwithstanding my enthusiasm, I defy anyone to make a direct link between carrying heavy logs across the countryside in order to build some bizarre construction and some aspect of real life. We can probably all agree that this represents a low-fidelity task. On the other hand, we can take a group out to an aircraft and design a task around an on-ground evacuation. Instantly we can see a higher level of fidelity even though the team processes involved in the log construction exercise and the aircraft evacuation may actually be the same.

There are some other structural elements we need to consider. First, we need to set a time for the exercise. The time available needs to be sufficient for the team to achieve the goal, of course, but we also need to bear in mind the return on the investment. As a vehicle for training delivery, practical exercises are time consuming. We need to balance the time allocated within the course programme against the learning value. That said, there are additional benefits accruing to the use of practical tasks. They break up the training day by providing a change of tempo. They engage the students in the learning process in an active rather than a passive way. They can actually be fun.

Thinking back to the 3-part structure of role-plays, we also need a cover story. The scenario provides a context within which the task takes place. The cover story, in effect, establishes the face validity of the exercise and, so, its acceptability to the group. However, the scenario does not directly contribute to the exercise. Finally, we need to have a clear understanding of how we plan to conduct the debriefing. Although I did not specifically include the debriefing as a component of a practical exercise, our efforts will serve no purpose unless the behaviours exhibited by the team are examined within a theoretical framework.

## *Desert Crash – A Group Decision-making Exercise*

Before proceeding, it will be useful to read the exercise description at the end of this chapter (pages 89-92). We can summarise the exercise as follows:

Briefing – full instructions issued to group but information spread across 2 different sheets of paper.
Task – to make a series of decisions as a group.
Time – 1 hour.
Cover Story – team lost in the desert and needs to find way to safety.

The idea for this game came from watching another exercise being used on a course for Tax Inspectors. Having observed the first couple of moves, it seemed to me that the exercise lacked any clear purpose and could be better engineered to focus on specific behaviour rather than, as it seemed, to be simply a vehicle for keeping the team entertained. What I wanted to do was present the group with a set of forced decision points.

Let us look at the game in action. Having briefed the group and presented the information sheets, the start-point for the game – the aircraft crash site – is revealed. From that point on, every move is a decision. What's more, decisions are made in a condition of extreme ambiguity caused by the lack of geographical information available to the group.

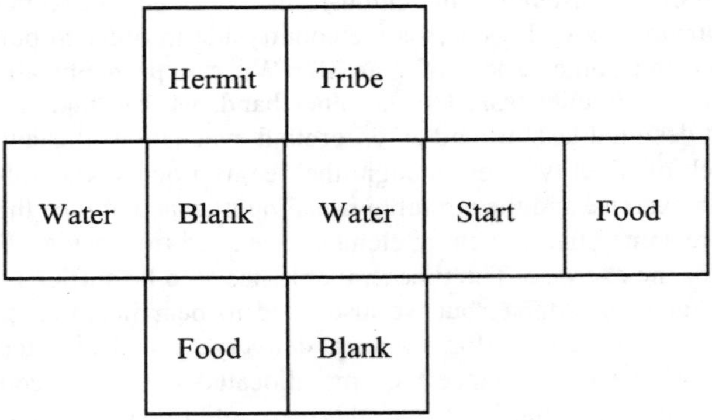

|  | Hermit | Tribe |  |  |
|---|---|---|---|---|
| Water | Blank | Water | Start | Food |
|  | Food | Blank |  |  |

**Figure 4.2 Desert Crash Opening Moves**

The game requires the team to decide on a direction of travel from the initial start point, which is an aircraft crash site in the desert. The direction of travel can be one of the 4 cardinal compass points. Travel is between points laid out in the desert on a geometrical grid. Most of the points contain items that will be of use to the team in reaching their destination, which is their base camp. The objective of the exercise is to reach base camp whilst meeting a number of criteria. Water must be found every 3 hours (which is the same as every third move), food is needed every 8 moves and an injured team member needs medical treatment within 24 moves. We have found that the exercise works best with groups of 5-8 players. We need a group of sufficient size to generate some differences of opinion but not so large that people can hide.

*How the Game Unfolds*

To understand the rationale of the exercise, I want to talk through the opening moves of the game. In Figure 4.2 I have shown the layout of the first 9 cards. Of course, the game players do not see the cards until they are laid out, one after another, as the team makes a move. So, the game opens with the 'Start' card on display. The team has to make a choice. The 'most-safe' choice will be determined by the criteria contained in the briefing. In reality, teams are surprisingly bad at analysing the situation and then applying the criteria as part of their decision-making process. We find that teams create explanations to justify decisions based on false information. Teams will change from safe to risky behaviour for no sound reason.

In our exercise, from the 'Start' position, the safest move is always towards water for the reason that this is the highest priority. Food is required after 8 moves whereas water is required every 3 moves. Teams who take food first often justify the decision on the grounds that they are bringing forward a task and therefore making life easier later on. Unfortunately, just as in real life, the teams do not know what will happen over time and so what appears to be a sound decision now could, in fact, be creating a problem later. So, the first behaviour we see is that of false application of priorities leading to increased risk. Once a team has moved 2 squares to the left, or west, we see the twin temptations of the hermit to the north and food to the south. In fact, neither square is attainable if the team wants to guarantee reaching water after 3 moves. Although the teams would know of the tribesmen to the north and the empty desert to the south of the first water square, they would have no other information about what lies adjacent to the hermit or food squares. At this point the talk usually turns to taking a gamble on finding water. Now we see attempts to rationalise choices using fairly flimsy reasoning. There are some pieces of known information: water is available to the east and west; water is needed every 3 hours; we cannot guarantee to get water of we move north or south. Even so, teams will take risky action with little or no discussion. As the exercise progresses further, teams will become even more risky in their selection of moves.

In this short description I have tried to illustrate how behaviour, which is reminiscent of CVR transcripts of crews trying to make decisions on an aircraft, can be enacted in a classroom situation. By getting a team to make real decisions, their performance is then available for analysis and discussion. We can compare actual performance against some notional model of decision-making and analyse why things go wrong. Of course, there are some who would dismiss such an approach. The exercise is artificial, the consequence of the decision is inconsequential, and humans can survive for more than 8 hours without food and so on – all of which is true. What we are trying to do is to generate ways of creating behaviour for subsequent analysis. Furthermore, we want that behaviour to be as close to the real thing, so to speak, as possible. As I said earlier, the players do not have to simulate decision-making; they have to do it for real. We needed a way to allow course members to develop, for example, their skills of decision-making and exercises like the one I have described provide just such an opportunity.

*Gate Game – A Simulation*

Whereas Desert Crash can be considered a 'toolkit' exercise, my second example was designed to get people to think about the nature of the aviation task, and so it falls into the 'lens' category. The trigger for the game was a week spent shadowing various jobs at Zurich airport. The inter-relationships between the various parts of an airport and the way in which problems seemed to reside as much in the spaces between functional units as within them proved irresistible. In simple terms, the exercise can be described as follows (see also pages 93-95):

> Briefing – group overview and then specific tasks issued to team members
> Task – to experience cross-functional working.
> Time – 30 minutes
> Cover Story – the team is responsible for preparing an aircraft for departure. The aircraft has to be loaded, the passengers checked and seats assigned, the pilot needs to calculate the fuel required and then the cabin crew have to supervise the boarding process.

The exercise is designed for teams of 4, one for each of the pilot, cabin crew, gate agent and ramp agent functions. Each of the players has an individual task to complete but each member of the team requires information from one or more of the other players. Because the tasks all require concentration, requests for information quickly become distractions and lead to increased stress. The tasks bear only a very superficial resemblance to the real world. To make the exercise more interesting, we switch roles as much as possible. So, the cabin crew play the pilots' role, pilots become gate agents and so on. To add pressure, additional tasks and problems are given to team members as the game progresses. As departure time approaches the passengers need to be boarded and we can measure the effectiveness of teams according to length of delays and the number of missing passengers.

The game works by creating a sufficiently complex individual task such that the player needs to focus on what they are doing. Next, the flow of work within the time available varies between players such that, at different times, some are idle while others are working hard. Third, the situation changes over time requiring constant communication between team players. Finally, no single player has all the information needed to complete the game. The most difficult aspect of the design process was making sure the calculations worked. Because I wanted the game to provide a measure of performance, it was important for successful teamwork to be observable. There are 3 criteria of success: number of passengers on board, on-time departure and aircraft weight. The effectiveness of the team will be reflected in all 3 numbers. I have included some examples of the materials used in the exercise at the end of the chapter.

*Building Exercises*

Now that we have looked at 2 examples of classroom exercises I want to outline the process of exercise construction. As I mentioned earlier, the inspiration for Gate Game came from watching the activity at an airport gate during an aircraft turn round. Translating this into a workable classroom activity required, first, the development of a workflow and then the creation of some tasks that could be accomplished in the time available. The 'workflow' is really the rough sequence of events. On a sheet of paper I mapped out who was supposed to be doing what and how the various task inter-related. The objective was to ensure that no individual could work in isolation. Of course, we have a contradiction here in that the essence of the game is to balance individual task completion with group collaboration. The 2 exercises I have described differ in the complexity of the workflow but each has a 'process' providing the spine of the activity.

The tasks assigned to the players vary enormously. In Desert Crash the 'task' is to reason out a course of action though talking. In Gate Game, the tasks are very superficial imitations of reality. The important point is that they require processing and, therefore, divert individual attention from the group task. The act of data processing, even if it is only adding up a list of numbers, disrupts group cohesion. So, having identified the spine of the exercise, we then need to generate activities for the syndicate members to complete. Activities will include processing words or numbers, developing a solution to a problem, discussing options or even constructing objects. The activities can be fragmented and distributed between syndicate members or they can be coherent and require all team members to participate.

Desktop exercises, like role-plays, are a form of low-fidelity simulation. The purpose of the exercise is to either provide an opportunity to rehearse skills or to allow critical focus to be applied to some aspect of the workplace. By simulating the task we are trying to generate insight in a way that is more effective than would have been achieved through an alternative teaching method. We will look at the use of higher-fidelity simulation within CRM later in this chapter but, next, I want to consider the use of questionnaires as a classroom activity.

## Questionnaires

Questionnaires are useful in that they can provide the class with feedback on their individual performance against some of the course goals. I want to make a distinction between the use of questionnaires in class and the use of psychometric tests. Many CRM courses make use of personality tests, the Myers-Briggs Type Indicator Test being widely used. To be used properly, commercially available psychometric tests should be administered and interpreted by qualified personnel. The use of psychometric tests in CRM training is beyond the scope of this book; my goal is simply to explore instruments we can easily use in class to focus attention on personal performance. Equally, the design of valid and reliable measurement tools is perhaps beyond the capability of most CRM training developers. So, we are really talking about activities that can serve as rough pointers towards a broad category of performance and no more. Despite that severe limitation, questionnaires should not be dismissed out of hand.

I have illustrated a range of questionnaires at the end of this chapter, each differing in its style. The first is based on the leadership work of Fred Fiedler and can be used in a discussion of leadership style. Although now very old, the Least Preferred Co-worker scale does allow individuals to consider where they sit on the task-team continuum. The response form (page 101-102) does require a little thought. It needs the candidate to consider a person, known to them but whom they would prefer not to work with given a choice. This is not the same as thinking about a person they dislike. The lowest score that can be gained on the test is 16. Typically, a score of 16 shows someone who has misunderstood the task. The respondent then evaluates their 'subject' on a set of scales.

Having used the questionnaire on a number of courses, we have been able to generate some norms for use in interpreting scores. Once the class has calculated their personal score, they need to see where their score falls on the following scale to see what is their preferred leadership style:

Pilots
```
     16------33|34------------50|51------56|57-------66|67--------------84|85-----93
     High Task|Moderate Task|Low Task|Low Team|Moderate Team |High Team
     36------40|41------------53|54------63|64-------67|68--------------76|77-----85
```
Cabin Crew

The second example of a questionnaire I have provided is a reworking of the venerable Holmes-Rahe Social Readjustment scale (page 103-104). Originally intended as a therapy tool for psychiatric patients returning to the community, the questionnaire is useful in that it gets people to confront life stress events, something you might discuss in a stress management workshop. This questionnaire asks respondents to indicate how many times they have encountered specific events presented as a list. Each event has a points value.

The idea behind the questionnaire is that repeated exposure to stress triggers is harmful to our health. Having completed the questionnaire, delegates need to check their scores against some benchmark. In this case, accumulated evidence shows that delegates with a score of between 0-150 have a 1 in 3 chance of being hospitalised with an illness in the next 12 months. In actuarial terms, this is the risk to which the whole population is exposed and so this can be considered a normal score. However, a score of 151-300 brings with it an increased probability of experiencing ill health in the next 12 months. The probability now rises to 1 in 2. For scores above 301 the probability in now 4 in 5. As a caution, we have found that students on new hire courses typically produce scores above 500, reflecting the turmoil in their lives at that stage. So, perhaps a questionnaire like the Holmes-Rahe Scale is better used on recurrent CRM courses?

Finally, I have provided a questionnaire complied from various sources, which looks at assertiveness (page 100). In this questionnaire, respondents are asked to rate themselves. I developed the scales to reflect individual preferences against the

3 behavioural styles included in the Assertiveness model. The questionnaire is designed purely to provoke self-reflection and does not meet the criteria required of a psychometric test.

Constructing your own questionnaires, assuming you are not attempting to develop a sound psychometric instrument, can be straightforward. First, clearly identify the quality or characteristic to you want to explore. Next, get colleagues to generate as many statements associated with the target characteristic. At this stage you simply want statements, not examples of good and bad. Using techniques we will be discussing the in the final section, get your colleagues to sort the statements into piles of related items. Finally, take the individual piles – which will form the basis of the scales in your questionnaire, and get colleagues to arrange the statements on a continuum from strong to weak, positive to negative or whatever distribution is most appropriate for the scale. This much-abbreviated approach lacks scientific rigour but offers a quick and simple way of generating tools that can be used to promote discussion in class. This book is not, after all, a research primer; it is a tool kit for dedicated facilitators.

## Line Oriented Flight Training (LOFT)

The final training method I want to examine is one that is more traditionally associated with flight simulators although the underpinning concepts have been used in the classroom for some time. LOFT has its roots in the rise of high fidelity flight simulators. The increased capability made users realise that aircraft flight profiles could be rehearsed in the simulator. Furthermore, the concept of training as being something other than repeated systems failures began to take hold. Finally, the simulator was accepted as a tool for training crew social skills as well as technical knowledge. The basic concept behind LOFT is that the training event replicates normal line operations and the crew is then allowed to work through the problem in much as they would be expected to work in real life.

Designing LOFT scenarios requires 2 vital pieces of information: a clear understanding of the behaviours to be trained and a firm grasp of how the company route structure exposes weaknesses in crew performance. In short, LOFT as training methodology is firmly in the toolkit camp. We are now looking at developing skills in a high-fidelity situation. Therefore, we need to be able to describe the desired behaviour in such a manner that we can explain to crews what is required, identify gaps in performance and negotiate the means to bridge any gaps revealed in training. Second, because we also want to guarantee transfer to line operations, it is important that the scenarios have sufficient face validity. One way to achieve this is to use representative route segments.

The framework we used to analyse case studies can also be applied to LOFT development in that precursors and triggers can all be used to shape the scenario. To this list we can add distractors; those aspects of line operations that get in the way of problem solving, such as calls from ATC. The real power of LOFT, however, should be found in the debriefing and this, in turn, puts great demands upon the skills of the facilitator.

We can exploit the basic idea of LOFT in a classroom environment and, by removing the need for high-technology, we can broaden the training group. For example, here is a section from an annual recurrent cabin crew Safety and Emergency Procedures course:

Objectives
> Demonstrate the use of Personal Breathing Equipment (PBE).
> Rehearse fire fighting procedures.
> Describe techniques to use in different fire situations (including on ground).
> Demonstrate teamwork skills.

Scenario
> Aircraft is now airborne in the cruise. The crew are in mid-cabin doing service. Crew become aware of a commotion in rear of aircraft, thick smoke can be seen coming from rear galley. Hot rolls are burning in the oven.
>
> Crew take initial actions. Use PBE.
>
> On reporting to the pilot, flight deck will apply maximum ventilation.

Post-incident review
> Debrief team. Explore decision-making and communication aspects of fire.

Tutorial
> Triangle of fire.
> Cover alternative fire locations.
> Do mandatory checks in post-incident reviews (PDIs).
> Refuelling during boarding.
> Review effects of stress on performance.

The exercise is undertaken in a classroom with course members playing the roles of the operating crew and the passengers. The classroom seats are arranged in rows as per a section of the aircraft. Demonstration safety equipment is positioned in the room in the same relative positions as if it were the aircraft. The facilitator will have created some additional problems to throw into the scenario as the 'performing' crew work through the drills. The rest of the class act as observers and, in this case, were provided with a set of CRM behavioural markers to use in the final discussion. The duty crew have to rehearse the appropriate drills as per company SOPs, demonstrate the use of various pieces of safety equipment but also employ CRM skills. Note that we have also build in a need to liaise with the pilots. Once the scenario has worked through to a conclusion, the observers, first, comment on the procedural aspects of the exercise and, then, discuss the CRM aspects using the markers as a discussion framework. The class is encouraged to share their experiences from the line to support key points or to offer possible alternative courses of action. The role of the facilitator is to guide the discussion and to ensure that all of the technical issues are discussed as required by regulations.

In this next example we look at how the model can be applied to a pilot initial course. The course is run over 2 days and is divided into 4 half-day modules. Each module starts with a desktop LOFT. We took a fairly typical but nonetheless challenging sector; one that new hire pilots are likely to encounter early on in their careers. We split the profile in to 4 chronological sections and these are used to drive the modules. Each module begins with the class having to solve a problem. The effectiveness of the solutions is then discussed, first, in the context of the relevant CRM syllabus topics and, then, within the line experience of the facilitators. In this way, the expertise of the facilitator is shared with the class.

SCENARIO PART 1

Divide class into pairs ('crews') and explain that they are going to undertake a flight from Bologna to Florence. Explain that all documentation is available.

Issue weather, NOTAMs, Flight Plan.

The clearance is for a Peretola 5L departure from Bologna Rwy 12 to Florence for an ILS to Rwy 05. Full aircraft and a slot time to meet. Wintertime. Florence weather, light snow, braking action medium.
Weather 1200' cloud base with viz 3500.

Explain the concept of risk and ask syndicates to list risks associated with route. List identified risks on the risk assessment form.

| Risks | Scale of Hazard |
|-------|-----------------|
| 1. | Low 1-2-3-4-5-6-7-8-9 High |
| 2. | Low 1-2-3-4-5-6-7-8-9 High |
| etc | Low 1-2-3-4-5-6-7-8-9 High |

Individually, evaluate risks on scale. Then get group to compare scores and give reasons to explain differences.

Review concept of risk and then identify 2 key CRM skills we want to look at next – Communication and Teamwork.

*Building a Classroom LOFT*

In a Classroom LOFT the spine of the exercise is formed by the procedures and checklists the teams will use. Similarly, the activities assigned to the 'players' are drawn from procedures. What we need in order to initiate the learning process is a contingency: some event that will disrupt the normal exercise of a drill. What we hope to do is use contingencies drawn from actual experience. In our planning, therefore, we need access to company event reporting systems or experienced crews who can identify operational issues that add risk to the specific procedure is not completed properly. Of course, the risk-inducing factor will be some property of the operational context. By exploiting operational knowledge, we can identify the optimum sequence of events within the procedural framework.

With our game plan to hand, the next step is collating all the resources needed to make the exercise work. Typically this will include all the paperwork normally available to crews on board the aircraft. We can include items of equipment if available and if it adds value to the exercise. We also need to be able to represent off-board agencies, such as ATC, if required and this may need some investigation if we are to develop a credible scenario.

The critical aspect of developing the LOFT is identifying the contingency that will cause the actors to have to deal with real problems but in a manner that makes the crew processes accessible. By 'accessible', I mean acted out in front of observers and open to analysis afterwards. Therefore, the problems need to be more than component failure followed by drill rehearsal. In your planning you will need to ensure that the failure creates a problem that needs crew skills in order to resolve.

## Conclusion

In this chapter we have laid out the process by which we begin to put courses together. We have established our training goals and explored the main activities we can use in the classroom. We have see what is involved in developing our own activities. In the next chapter we will look specifically at the delivery of instruction.

The inspiration for an activity can come from something as simple as watching a children's game or reading an incident report. However, the flash of inspiration is only the start. The real effort comes in turning the idea into something that delivers results. The simplest of activities can deliver outstanding results and the smartest of activities can fail spectacularly. Never underestimate the inability of your customers to see the point of an exercise. I once ran a course on conflict resolution and the heart of the session was a practical exercise based on buying a replacement aircraft. The exercise generated conflict between the syndicates and the rest of the course explored the techniques of resolution. The client didn't like the exercise because none of the delegates could agree on the correct choice. I rest my case.

## What Next?

Think about a specific skill that, in your experience, needs further development in your company. Identify a range of possible training activities that could be used to develop that skill in training. Compare the various methods and, having selected the most preferable, spend 30 minutes making notes about what that preferred training method would involve. Once you have sketched out your ideas, put them in an envelope and leave them in a draw for a few weeks. When you return to them, see if a solution is now more apparent. Good ideas, like large mammals, have particularly long gestation periods!

## Useful Websites

A good starting point for finding information on the web is: www.aviation-safety.net

The following English-speaking sites are a rich source of accident and incident reports:

US National Transportation Safety Board    www.ntsb.gov
UK Air Accident Investigation Board    www.aaib.dft.gov.uk
Australian Transportation Safety Board    www.atsb.gov.au/aviation
Transportation Safety Board of Canada    www.tsb.gc.ca

Here are 2 academic journals that warrant the occasional inspection:

*International Journal of Aviation Psychology* (Lawrence Erlbaum Associates)
*Human Factors and Aerospace Safety* (Ashgate)

Finally, you can find examples of classroom materials at www.turboteams.com

Currency adds to credibility and is the responsibility of all facilitators to keeping touch with their field.

## References

FAA, (2004) AC120-51E Crew Resource Management Training.

# Instructions for running 'Desert Crash' Exercise

Give the briefing sheet to one team member and give the separate key to cairn symbols to a different group member.

Place the start point (square 1) on the table.

Delegates have 5 minutes to read and absorb information. After 5 minutes, ask the team in which direction they are going to walk. Place the appropriate square in position when they have made their choice. Each move is completed once the team chooses a direction of travel (north, south, east or west) and the appropriate map square is put in place.

If the team lands on any of the hermit sites (squares 6, 23, 25, 28), the information to be given is 'not all tribesmen are dangerous'. Visits to the tribesmen in square 4 are dangerous, though, and this move will result in the team's destruction.

No additional information is to be provided (other than when landing on a hermit site), such as the approximate distance from base camp or the nature of the response to be expected from local tribesmen.

Teams that violate one of the rules (food, water) lose a life and restart the game at the last square in which they were 'safe'.

The numbered grid below is to enable the instructor to easily identify the appropriate map squares. Copy the grid plan to use during the exercise. A marker, such as a coin or piece of Bluetak, will help you to keep track of moves. Before you start the game, remember to write the numbers on the backs of the squares!

|    |    | 8  | 7  | 6  | 4   |     |    |    |
|----|----|----|----|----|-----|-----|----|----|
|    |    | 10 | 9  | 5  | 3   | 1*  | 2  |    |
|    |    | 17 | 16 | 15 | 14  | 13  | 12 | 11 |
|    |    | 24 | 23 | 22 | 21  | 20  | 19 | 18 |
|    | 30 | 29 | 28 | 27 | 26  | 25  |    |    |
|    |    | 34 | 33 | 32 | 31* |     |    |    |
|    |    | 39 | 38 | 37 | 36  | 35  |    |    |

*Square 1 = start, Square 31 = home

# Desert Crash

Instruction Sheet

You are a group of archaeologists flying out to join a project in the desert. The project team is investigating an ancient civilisation that once lived in this area.

You were flying to the project headquarters, which is in an old desert fort, but your aircraft has crashed. Although you all managed to get out safely, the pilot was very badly injured and all of your supplies were lost. You need to get the pilot to the fort where there is a doctor. The pilot may not survive another day.

The people who lived in this area built a network of marker beacons to aid the nomadic tribes. Each beacon is a point on a grid. The grid is aligned with magnetic north and each cairn is an hour's walk from its nearest neighbour.

Many of the markers are supplied with food or water. Hermits who can give information to travellers have occupied some of them. Some of the markers are in areas where the local tribesman cannot be trusted. Each marker bears an inscription that tells you what may be found at each of the neighbouring locations.

Because of the heat you must find water every 3 hours. Its been some time since you last ate and you will also need to find food every 8 hours or else you will probably be too weak to continue.

As a group, you need to get the pilot and yourselves to safety. Stick together at all times.

Once you have made your group decision about where to move next, the instructor will give you the next card in the sequence.

*Expedition Base Camp*
*Desert Fort*

*Dear Joe,*

*Hey, it's good to hear that you got that place on the team! You'll love it out here - although it's pretty hot. Still, you always said you look better with a tan! The ancients who set this place up really knew how to make sure life went on. They set up these beacons, piles of stones with pictures to tell you what was available around you.*

*A bag means food and a thing like a flask means that water can be found. If the way ahead is impassable then there is a symbol that looks like rocks. A skull means watch out for the locals. They also have this thing which looks like a guy carrying a spear. This seemed to be the symbol for a hermit. These wandering tribesmen believed that these guys had information that could get you out of a dangerous situation.*

*We're starting to excavate some of the beacons to look for objects but the job is made difficult because they are still used by tribesmen - can you believe that! Caravans regularly cross the desert and it seems that, if you use any of the supplies, then you are obliged to replenish the system on your next visit. Great, eh! The system's been going for hundreds of years. Anyway, I've got to get off as I'm travelling to some of the northern beacons. I look forward to seeing you soon,*

*Jane xxx*

# Debriefing 'Desert Crash'

Look for the following points, which may be used as prompts for discussion afterwards;

- How were decisions made?
    - How were alternative proposals dealt with?
    - voting, majority view, agreed consensus?
- Were unfounded assumptions made (such as that the cairns followed a predictable pattern; fort was to the south on the basis of Jane's letter)?
- Did everyone have a chance to express their point of view?
- Did roles emerge within the group?
    - supporting, proposing, challenging?
    - often, deference is paid to the most senior pilot.
- On what basis were these roles assigned?
    - informal adoption of roles, such as keeping track of need for food and water.
- Did everyone understand what was going on?
- How did people deal with uncertainty?
    - were decisions put off until more information was available?
- For those groups which failed to meet the objective, how did the group feel?
- Did anyone suggest breaking the rules in order to achieve the goals?
- What were the team members' levels of involvement?
    - who spoke most?
    - most junior members usually most quiet, even when they can see things going wrong!
- How were the decisions made in terms of the risk which applied at each stage?

# Gate Game Team Briefing Sheet

You are part of a team responsible for preparing an aircraft for departure. You will be given further information about your specific responsibilities. During the exercise, your group tutor will act the part of any outside agency you may need to deal with. Your company has published a planned timetable of events that you should aim to follow, together with the approximate time before departure, in minutes, that each activity should take place.

| | | |
|---|---|---|
| Cleaning | started by | -30 |
| | completed by | -20 |
| | | |
| Catering loading | started by | -30 |
| | completed by | -20 |
| | | |
| Fuelling | started by | -20 |
| | completed by | -8 |
| | | |
| Catering/cabin equipment | check completed | -15 |
| | | |
| Passenger Boarding | starts | -20 |
| | completed | -5 |
| | | |
| Take-off Weight | confirmed | -3 |
| | | |
| Doors close | | -3 |
| | | |
| Pushback tractor | connected | -3 |
| | | |
| Depart | | On Time |

You have 30 minutes to complete the task; the clock starts when your individual tasks are issued. Any Questions?

# Gate Game Master Timetable

Hand out Team Briefing sheet. Handle any questions about the nature of the exercise.

Nominate time to be away from the gate (30 minutes after issue of task sheets).

Start Time.　　　Hand out specific task sheets to each role. State that the cleaners are aboard the aircraft and the caterers are loading meals at the moment. Tell the pilot that refuelling has just started and that 9,500 kgs will be put in the tanks. The pilot will need to notify the refuelers of any variation in that amount. Also, during walk-round, First Officer noticed low engine oil contents on one engine. Ground Maintenance have been informed.
Issue Task Card Pilot 1 - Oil Replenishment
Issue Task Card Gate 1 - Details of Passengers bumped from Previous Flight

+5 mins  Individual team members should have started tasks
　　　　　　　Pilot - calculate additional fuel required
　　　　　　　Gate - allocate passengers, determine empty seats
　　　　　　　Cabin - calculate meals
　　　　　　　Ramp - check palettes

Issue Task Card Gate 2 - Details of Standby Passengers
+7 mins  Gate Task 1(details of connecting flight passengers)
+10 mins　　　Boarding should have started according to timetable.
+11 mins　　　Cleaners leave the aircraft (tutor to inform cabin)
Boarding can now commence (cabin or gate should initiate boarding) (Late confirmed passenger)
Cabin should demand additional meal trays, meals and special meals

+12 mins　　　Ramp Task (request for priority freight)
　　　　　　　Gate Task 4 (late confirmed passenger)
(Boarding+3　　Issue first set of passengers to Cabin [Cabin 1])

# Gate Game Pilot Briefing

Your job is to calculate the fuel required for the flight and then, when all of the information is available, calculate the take-off weight for the aircraft.

Refuelling has just started and the fuel bowser operator has been told by flight operations to put 9,500 kg into the aircraft initially. You will need to decide on the exact amount of fuel you require and tell the operator.

You will need to consider the following limitations when making your calculations:

Ops empty weight (this is the basic weight of the aircraft) =27,515 kg
Max Zero fuel weight (this is the maximum weight of the aircraft once it is loaded with passengers and cargo but without any fuel)        =43,091 Kg
Max take-off weight  (this is the maximum weight of the aircraft at take-off, including fuel)      =52,390 Kg

Fuel Calculation

Fuel required = (distance divided by groundspeed x fuel consumption) + taxi fuel + diversion fuel + reserve fuel.
Groundspeed = True Air Speed + tailwind (or - headwind )

Route: startpoint to funland = 1045 nm

Average fuel consumption =        3,500 Kg/hr

True air speed              462 kts
Headwind component      44 kts

Groundspeed =  _____?

Fuel required......

# And just for fun … A leadership exercise!

Time Required – 45 minutes

Materials
        1 raw egg
        Small roll of sellotape
        40 Plastic drinking straws
        Marking pens
        Spare eggs
        Copies of the Team Briefing Sheet for each group.
        String
        4 sheets of A4 paper per team

## PROCEDURE

Divide the class into teams of 4-5. Tell the groups that they represent companies that produce crew safety devices. They are all working on prototypes for a system to be installed in a new aircraft project.

Each group has the task of designing, constructing and evaluating their crash protection device. Give them 30 minutes to design and construct their system using the material supplied.

At the end of the construction phase, tell the groups that there will be an evaluation flight. Should the eggs break during the test flight the company will lose the contract (and have to clean up the mess!). Each team can undertake up to 3 pre-trial 1-metre test flights.

The evaluation will be a 2.5-metre test flight (facilitator stands on desk and drops device!).

## DISCUSSION POINTS

1. What was the problem?
2. How was it broken down?
3. Who did what?
4. What roles emerged?
5. How were any disagreements dealt with?
6. Did any teams make use of the pre-trial evaluation flights?

# Team Briefing

Your team represents a company that designs crash protection devices for aircraft. You are working on an idea using new materials and which has not been properly tested yet.

For this exercise you will have 30 minutes to design your crew crash protection system. On completion of the design and construction your system will be required to undergo flight testing. We will be using eggs instead of crash test dummies. If the egg breaks, you lose the contract.

Before the final testing phase, you will be permitted to make up to 3 test flights from a height of no more than 1-metre. You must seek authorisation before completing any test flights.

The competition test flight will be from a height of 2.5-metres.

## Leadership Style Questionnaire

Think of all the people you have ever worked with and then think of one specific person with whom you could work *least well*. It may be someone you work with now or someone you may have worked with in the past. It does not mean that the person is someone you *do not like*, but just someone with whom you had difficulty getting the job done.

Keep a picture of that person in your mind as you now describe them on the following scales. Each scale consists of a pair of words which are opposite in their meaning, such as 'very neat' and 'very untidy'. Between each pair are 8 spaces to form a scale like this;

Very Neat:___:___:___:___:___:___:___:___:Very Untidy
       8    7    6    5    4    3    2    1

Now, if you reckon that the person you have in mind, the one with whom you work least well, is usually 'quite neat' you would a mark in the space marked 7, like this;

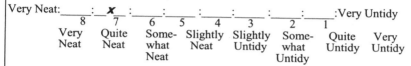

Very Neat:___: **X** :___:___:___:___:___:___:Very Untidy
     8    7    6    5    4    3    2    1

| Very Neat | Quite Neat | Some-what Neat | Slightly Neat | Slightly Untidy | Some-what Untidy | Quite Untidy | Very Untidy |

If you ordinarily think of the person as only being 'slightly neat' then you would put a mark in space 5.

Look at the words at both ends of the line before you make your mark on the scale. *There are no right or wrong answers*. Work rapidly: your first response is likely to be your best. Please do not leave out any of the items and only mark each scale once.

Now, describe the person with whom you can work *least well* on the scales below.

| | | | | | | | | |
|---|---|---|---|---|---|---|---|---|
| Pleasant | : | : | : | : | : | : | : | : Unpleasant |
| Friendly | : | : | : | : | : | : | : | : Unfriendly |
| Rejecting | : | : | : | : | : | : | : | : Accepting |
| Helpful | : | : | : | : | : | : | : | : Unhelpful |
| Unenthusiastic | : | : | : | : | : | : | : | : Enthusiastic |
| Tense | : | : | : | : | : | : | : | : Relaxed |
| Distant | : | : | : | : | : | : | : | : Close |
| Cold | : | : | : | : | : | : | : | : Warm |
| Cooperative | : | : | : | : | : | : | : | : Uncooperative |
| Supportive | : | : | : | : | : | : | : | : Hostile |
| Boring | : | : | : | : | : | : | : | : Interesting |
| Quarrelsome | : | : | : | : | : | : | : | : Harmonious |
| Self-assured | : | : | : | : | : | : | : | : Hesitant |
| Efficient | : | : | : | : | : | : | : | : Inefficient |
| Gloomy | : | : | : | : | : | : | : | : Cheerful |
| Open | : | : | : | : | : | : | : | : Guarded |

**Life Stress Events**

| Event | ①How many times in last 12 months? | ②Multiply by... | | ③Put result here... |
|---|---|---|---|---|
| Death of Spouse or partner | _____ | x100 | = | _____ |
| Divorce | _____ | x75 | = | _____ |
| Marital Separation | _____ | x65 | = | _____ |
| Death of a close family member | _____ | x63 | = | _____ |
| Personal injury or illness | _____ | x53 | = | _____ |
| Marriage | _____ | x50 | = | _____ |
| Dismissal from work | _____ | x47 | = | _____ |
| Marital reconciliation | _____ | x45 | = | _____ |
| Change in health of family member | _____ | x44 | = | _____ |
| Pregnancy | _____ | x40 | = | _____ |
| Lost interest in sex | _____ | x39 | = | _____ |
| Gain of new family member | _____ | x39 | = | _____ |
| Change in personal financial situation | _____ | x38 | = | _____ |
| Death of a close friend | _____ | x37 | = | _____ |
| Change to a different line of work | _____ | x36 | = | _____ |
| Having arguments with spouse or partner | _____ | x35 | = | _____ |
| Take out a major loan on house | _____ | x31 | = | _____ |
| Bank threatens to call in mortgage or loan | _____ | x30 | = | _____ |
| Change in responsibilities at work | _____ | x29 | = | _____ |
| Trouble with the in-laws. | _____ | x29 | = | _____ |
| Outstanding personal achievement | _____ | x28 | = | _____ |
| Partner begins or stops work | _____ | x26 | = | _____ |
| Begin or end a training course | _____ | x26 | = | _____ |
| Change in living conditions | _____ | x25 | = | _____ |
| Trouble with boss | _____ | x23 | = | _____ |

Change in work
  hours or conditions   _____    x20    =    _____

Move home    _____    x20    =    _____

Take up a new sport
  or stop an old one    _____    x19    =    _____

Meet new friends
  or lose old ones.    _____    x18    =    _____

Taking out a
  small bank loan    _____    x17    =    _____

Change in sleeping habits    _____    x16    =    _____

Trying to diet    _____    x15    =    _____

Taking a holiday    _____    x13    =    _____

Christmas    _____    x12    =    _____

Speeding tickets or
Parking fines.    _____    x11    =    _____

④Now add up the total    _____

(based on Holmes-Rahe Social Readjustment Scale)

**Interaction Style**

Read each of the following statements and ask yourself if you tend to agree or to disagree with each one. Mark each of the statements accordingly. So, if you reckon that, generally speaking, you would tend to agree with a statement, then circle the number of the question. Work quickly through the statements. Remember, there are no right or wrong answers and your immediate reaction to each statement is probably the most accurate reflection of your true feeling.

1. The best way of handling conflicts is to avoid them.
2. I never assume that I will get what I want in life.
3. I find that its best not to give people a choice.
4. I do value the contribution of others.
5. I usually check my assumptions.
6. I do not like to offend people.
7. Conflict can be a creative force.
8. In this world there are winners and losers.
9. I usually go along with what people want.
10. I always tell people exactly what I think of things.
11. I rarely give praise because its not often deserved.
12. I am happy to take responsibility for things.
13. You do not need to offend people in order to get the job done.
14. I do have a tendency to jump to conclusions.
15. I like to get the views of others.
16. If people take offense then it is not my problem.
17. I usually tell people exactly what I think of things.
18. In groups, I usually keep my views to myself.
19. I like to discuss ideas.
20. I usually let others take responsibility for things.
21. I am happy just to do what I am asked.
22. I often get angry or upset when I have a conflict with others.
23. I try to get the point of view of those with whom I am having a conflict.
24. It is important not to upset people.
25. I always try to get others to agree with me when we have a decision to make.
26. I like a quite life.
27. I consult others before deciding on matters that affect them.
28. I know I am not as constructive as I could be when I disagree with someone.
29. I try to make sure that everyone gains something from a dispute.
30. I find compromise is usually the best solution.
31. I play to win.
32. I accommodate myself to the other persons view.
33. We openly integrate ideas of both persons.
34. I often find ways to accept the other persons solution.

35. I don't like to give in until I get what I am after.
36. I fully express my ideas and feelings and urge the other person to do the same.
37. I push to have my approach or ideas prevail.
38. We find solutions that take both our views fully into account.
39. I am happy to try solutions proposed by the other person.
40. I like to get every-ones' concerns out in the open and we problem-solve together.
41. I don't usually resist the views expressed by the other person.
42. I usually get my ideas accepted.

**Marking Frame**

Circle the numbers of the questions you tended to agree with, i.e. those you circled on the question sheet. Then, add up the number of questions in each of the 3 columns you agreed with. You should end up with a score against the aggressive (ag), assertive (as) and non-assertive (na) interaction styles. The highest number indicates your prevalent style.

| | | |
|---|---|---|
| 1 | 4 | 3 |
| 2 | 5 | 8 |
| 6 | 7 | 10 |
| 9 | 12 | 11 |
| 18 | 13 | 14 |
| 20 | 15 | 16 |
| 21 | 19 | 17 |
| 24 | 23 | 22 |
| 26 | 27 | 25 |
| 30 | 29 | 28 |
| 32 | 33 | 31 |
| 34 | 36 | 35 |
| 39 | 38 | 37 |
| 41 | 40 | 42 |
| na ___ | as ___ | ag ___ |

# Example Training Objectives

*Course Goals*

- To introduce delegates to the nature of the aviation task.
- To introduce delegates to aspects of personal performance which impact on the aviation task.
- To introduce delegates to aspects of group performance which impact on the aviation task.
- To introduce delegates to aspects of organisations which impact on the aviation task.

## Supporting Objectives

The Aviation System.
1.1  To discover the nature of work processes in aviation.
1.2  To discover aspects of the collaborative nature of aviation.

Human Information Processing.
2.1 To describe the processes of perception.
2.2  To describe the function of attention.
2.3  To describe the constraints on working memory.
2.4  To describe how information is stored in long-term memory.
2.5  To describe how limitations in information processing lead to errors.
2.6  To describe the role of information processing in Situational Awareness.

Human Error.
3.1 To describe what is meant by error.
3.2 To describe three categories of error.
3.3 To identify the causes of error.
3.4 To identify factors which support violations.
3.5 To identify the impact of automation on error.
3.6 To identify error detection mechanisms.
3.7 To discuss the role of Inquiry and Critique in error reduction.

Stress.
4.1 To describe the difference between arousal and stress.
4.2 To identify the causes of stress.
4.3 To illustrate the physical, behavioural and emotional effects of stress.
4.4 To describe the management of stress.
4.5 To describe the difference between stress and fatigue.
4.6 To describe the effect of stress and fatigue on vigilance.

Personality, Attitudes and Motivation.
5.1 To describe what is meant by personality.
5.2 To describe what is meant by attitude.
5.3 To outline the origins of predictable behaviour.
5.4 To illustrate the effects of personality and attitude at work.
5.5 To develop a simple model of self-esteem.
5.6 To identify the relationship between self-esteem and the motivation to act.
5.7 To identify models of motivation.

5.8 To consider the effects of motivation on individual performance.

The Structure of Groups.
6.1 To describe the concept of team roles.
6.2 To identify elements of the task which support group functioning.
6.3 To describe how group norms emerge.
6.4 To describe the interplay between tasks, roles and norms.
6.5 To discuss the emergence of structure in groups.

Leadership and Teamwork
7.1 To differentiate between task and team leader styles.
7.2 To describe sources of leader power.
7.3 To describe effects of status within groups.
7.4 To discuss the role of delegation in task management.

Communication.
8.1 To identify types of communication.
8.2 To describe a model of communication.
8.3 To identify the functions of communication.
8.4 To identify barriers to effective communication.
8.5 To discuss advocacy and feedback as modes of communication.
8.6 To identify the role of communication in conflict resolution.

Decision Making.
9.1 To identify the processes by which decisions are made.
9.2 To identify the effect of group processes on decision-making.
9.3 To identify the barriers to effective decision-making.
9.4 To identify methods of improving decision-making.
9.5 To discuss conflict resolution as a process within decision-making.
9.6 To discuss workload management strategies as decision-making.

Organisational Error
10.1 To identify causes of latent and active error.
10.2 To discuss organisational elements in a model of accident causation.
10.3 To consider the implications of SOP design in promoting safe operations.
10.4 To discuss the development of a company safety culture.

Culture
11.1 To identify the nature of culture.
11.2 To identify the origins of culture.
11.3 To identify the effects of culture on performance.

Assertiveness
12.1 To distinguish between assertive, aggressive and non-assertive behaviour.
12.2 To identify delegates' preferred interaction style.
12.3 To examine the effectiveness of alternative interaction styles.

**Example Lesson Planning Sheet**

*Topic:* Workload Management

*Linked Topics*: Information Processing, Communication, Situational Awareness.

*Do I need this to understand another topic?*


**Syllabus Reference**

JAR-OPS: ?
FAA: ?


**Sources**

Note down reference documents, suitable case studies etc.

*What does it mean for us?*

From you investigations, is this topic especially significant within you operation? Is there any particular sector where, in this case, workload management is critical? Seek the views of the Safety Dept, Fleet Managers and Line Trainers.

*Do I have an example from my airline?*

Gather background examples from within your own airline. You might not use these in your training might it might be useful to have them as a guide to selecting similar events from other airlines.

*Examples from other airlines*

Gather relevant examples from other sources.

# Chapter 5

# Delivering Training

## Introduction

In this chapter I want to explore the skills and techniques required for the effective delivery of training. We will concentrate on a classroom situation but the content of this chapter applies equally to all training situations. Remember; facilitation is simply the process of assisting someone to learn. If you are working in a one-to-one situation, as in a simulator, or in a large group situation, as in a classroom, the skills are the same. I want, first of all, to look at how we plan a training session and then I will move on to presenting information to the audience. The key instructor competences covered in the section are:

> B1.1 Clarifies CRM in the context of overall training environment
> B2.2 Clarifies training objectives and methods
> B2.3 Ascertains and supports learners' needs
> B3.1 Clear and persuasive communicator
> B3.4 Organised, systematic lesson plans
> B3.6 Actively clarifies understanding with learners
> B3.5 Clear, accurate presentation materials
> B4.1 Uses exercises and activities designed to maximise CRM training objectives

## Preparing for the Classroom

Before proceeding, let us take stock. In the previous chapter we developed our training objectives and we identified the media we wanted to use in order to achieve those objectives. We started gathering information on our Lesson Planning sheet. We considered how best to sequence the training elements we intend to use in our module of instruction. In this chapter I want to concentrate on planning the 'lesson' unit of our module. A useful structure to have in mind while compiling training sessions is: Introduction, Body and Conclusion.

### The Introduction

The Introduction allows us to set the scene and lay down the ground rules for what is to come. A handy mnemonic to use for planning the Introduction is INTROSH and its use is illustrated in Table 5.1. It is important to remember that the

Introduction is a structuring device. No actual learning is occurring; the students are simply being prepared to learn. The introduction need only take a minute and certainly should not take more than 10% of the planned classroom time. Of course, an introduction element is applicable to all types of training activity and not just formal classroom sessions.

**Table 5.1 The Introduction**

| Element | What it is | What I'd Say (for example) |
|---|---|---|
| Introduction | Link | "In this next session we are going to explore Workload Management". |
| Need | Justification | "Failure to schedule tasks adequately can increase the risk of things going wrong later in the flight". |
| Title | Signpost | Show a Title Slide. |
| Review | Integration | "In a previous session we looked at human memory can be fallible.  What part does memory play in workload management?" |
| Objectives | Signpost | "Our objectives for this session are …".  Show Objectives or broadly describe them. |
| Scope | Preparation | "In this session I want to build on our previous examination of the case study and fit some of our conclusions into a theoretical framework". |
| Handout | Administration | "There is a handout covering the key points that I will give out at the end of the day". or "You may want to take notes as you will be putting the theory into practice in the next session". |

*The Body*

In the body of the session we do most of the work of facilitation.  One easy way to structure this phase is to use your objectives as a framework.  From Table 4.1 I have extracted the objectives I consider most appropriate to my lesson.  I have reordered them slightly, relegating 1 of them to the status of sub-objectives:

Describe the techniques available to manage workload.

Describe the $x$ common failures in checklist application.

Describe the purpose of checklists and procedures.

Describe $y$ techniques for ensuring safe application of checklists.

I can now distribute my resource material between my objectives. At this point I need to start thinking about how I will present the material. I need to think about the time I have available and how I am going to allocate it to the 3 blocks I have identified. From my initial inspection I feel that the first objective will not require too much time and, perhaps, objective 2 will require the major proportion of the time. In some ways, objective 3 looks like it could lend itself to acting as a review vehicle for the discussion engendered by the previous 2 objectives.

We need to think about our role as a facilitator and should be looking for ways to structure our material so that it will enhance the students' learning. In order to do that we need to consider the key events of instruction outlines in Table 5.2.

## Table 5.2 The Events of Instruction

| Event | What it does | What I might say |
|---|---|---|
| Preparation | Points the way, signposts, provides scaffolding to judge progress through the material | "We have just seen that …". "So, to summarise …". "For the next 10 minutes I just want to …". "You remember when we looked at, well now …". |
| Focusing Attention | Signifies what is important | "Note how the design of this checklist …". "Compare these 2 SOPs for the same task". "What is our SOP for…?." |
| Enhancing Storage | Assists learning through better understanding | "So, to summarise …". "A good way of thinking about this would be …". "It's a bit like when you …". |
| Assisting Recall | Eases remembering | "Let me repeat, a good SOP is …". "Remember INTROSH and you will always get the first bit of your lesson right". |
| Promoting Transfer | Clarifies Relevance, Reinforces the need to know the content of the session | "The SOP is designed this way because …". "Why do you think the we do it that way?" What happened when you last did …?." "How did you get around that problem?" |
| Monitoring Performance | Allows students to gauge their own progress in terms of grasping the content | "What were the 3 main weaknesses in checklist design?" "Which of these 2 SOPs is better? Why?" "If I wanted to write a procedure for $x$ how would I do it?" |

In both the lecture and the lesson formats we have decided to engage the student in a form of debate. Through the use of words, supported by visual aids where appropriate, we plan to achieve a level of understanding of the topic under review. We apply the structure afforded by the Events of Instruction to marshal a set of statements. Each statement contributes to a body of evidence that, at the end of the session, promotes a degree of understanding of the topic in the individual students. Some of the statements we will offer to the class for their consideration. Some of the statements we will elicit from the class. Through questioning, we will get the students to analyse examples we may offer, interpret their own experiences or construct hypotheses based on the content of the session. The lecture mode of delivery allows less engagement with the class, which is the main reason why it is not recommended for CRM classes.

## The Conclusion

An old adage of public speaking is 'Tell them what you are going to tell them, tell them, then tell them what you just told them'. In the conclusion, we tell them what we just told them. Again, the conclusion should not exceed 10% of the total classroom time. First, briefly review the key points of the session (or get the class to do this). You can refresh the class' memory of the objectives while this is going on. Ask the class if there are any questions or points people want to raise. You may want to pose a couple of questions to the class at this stage as a check of understanding. Finally, deal with the housekeeping. Tell them what is coming next, if there is a handout to collect, what time you want them back from the break and so on.

## Finishing Touches

### Step 5 – Developing Lesson Plans

Having finished our planning we now need to capture our thoughts on paper. We said at the start of the previous chapter that various documents exist to support the management of training. We also said that, especially during the first few courses, the facilitator would need some form of framework to direct their activity. This is traditionally been done through the lesson plan. A lesson plan is simply an aide memoire. It is personal to the instructor. The guiding principle should be that, which ever format you chose, it must be easy to use and readily understood in class. Nothing is guaranteed to destroy your credibility faster than losing your place in the script. An example lesson plan for a lesson on Workload Mangement is given at the end of this chapter.

### Step 6 – Preparing Visual Aids

Some consider Microsoft to be guilty of many things but probably by far the worst crime they have committed was to release Powerpoint onto an unsuspecting world.

Things were bad enough when the Overhead Projector (OHP) was the tool of choice but the ease with which brain atrophy can be induced has increased exponentially with the application of technology. But here is the paradox: on the one hand, *not* using Powerpoint is considered to be 'backward' and yet, the same people who advocate its use then complain about how dull presentations can be. OK – diatribe over (but, for a more reasoned debate see the work of Edward Tufte). The point in all of this – and the same argument can be levelled at OHPs – is that we forget that visual aids are simply that: visual aids to promote learning. What we actually see are 'visual verbals'; the facilitator projects a script onto the wall. Without a doubt, it makes sense to reveal the title of the session and there are some textual elements that can be reinforced through projection. It is good practice to declare the lesson objectives at the start of the session and this can be done with a projected image. However, it is important to remember that we can read off a page faster than we can say the words out loud. If you are going to read off the screen, the class will be at the end of the page well before you!

A good visual aid will undoubtedly facilitate learning. Visual aids allow us to bring the outside world into the classroom. Objects that are either too large to fit into our training venue or are too remote can be brought into view. Typically, a good visual aid will render comprehensible that which would take too long, or would be too difficult, to describe. Right away we can see that we are looking to transpose information from one format to another and visualisation is an art in itself.

Visual aids fall into 2 main categories. On the one hand we have those devices that project images while on the other we have devices to capture data. The first category includes TV, video, data projectors, computers, OHPs etc. The use of these devices triggers a host of technical and logistical questions. For example, what equipment is needed, what space are you using, does the shape of the room permit adequate visibility, what light level is required and so on.

My data capture category is just a fancy way of saying 'something you write on'. By this I mean chalkboards, whiteboards and flip charts. Here, again, issues of classroom layout visibility need to be considered. Although this is clearly the low-tech end of the spectrum, quite often a flip-chart is all that is available within the training accommodation. The flip-chart has many strengths, being simple, flexible and does not require batteries. Visual aids are important but choices should be driven by the needs of the student to understand concepts. Whatever aid you select, you must always consider the impact of clarity, legibility, visibility and use of colour; all aspects of a visual aid that will cause it to fail in its purpose of overlooked.

## Step 7 – Developing Handouts

The question of classroom handouts is a thorny one. There is an expectation on the part of many students that they will walk away with something in their hands. Some regulatory authorities seem to expect that a course manual will be produced. Of course, the real question is why would you want to hand out paper to students?

If you are using text-based case studies or questionnaires, then it may be administratively more convenient to produce a collection of papers that can be given out in advance of the course to allow delegates time to prepare. Both of these activities eat up time in class and so, by allowing the class to arrive already primed, we can speed things along. You may decide that, because classroom time is insufficient, extra background reading is needed to provide the appropriate depth of coverage. At the very least, you may want to provide a summary of the key points of each session. Whatever your motive, handouts need to be planned and they take time to prepare.

*Step 8 – Checking Facilities*

I was running a class in Africa during the rainy season. The classroom was a wooden hut with a tin roof. Whenever it rained it was impossible to make myself heard beyond the first row of seats. We were working 6 days a week but the administrative staff only worked Monday to Friday. So, on the first Saturday we were locked out. Fortunately, one of the First Officers was an ex-Boy Scout and, using his Swiss Army Knife we were able to dismantle a small toilet window. The second advantage of this individual was that he was small enough for us to then slide through the window so that he could open the doors from the inside. The moral of this story is if it can go wrong it undoubtedly will. You will arrive at a venue to find no power supplies for video players, monitors that are really old television sets that have seen better days, flipchart easels with no paper, marker pens that have dried up ... shall I continue? To make matters worse, clients are not always understanding or sympathetic to your problems. I was running an interview techniques course for a major food producer in the UK using their on-site training facility located next to the liquid gas delivery yard. Every morning a tanker would turn up and pump for 2 hours. The noise made working in the classroom impossible. Now, you can imagine that an interview techniques class involves a degree of talking. When I pointed out the challenges of trying to rehearse interviews in such an environment I was simply told to get on with it, after all, I was getting paid, wasn't I?

The question of facilities can influence your approach to course design. I once planned a highly interactive course based around the concept of Situational Awareness. I knew the operator had a large empty hangar next to the training room, ideal for what I had in mind. Only when I handed off the materials was I told that, in fact, the training would be delivered in a classroom at a different site completely. The whole course had to be reworked.

**Delivering Instruction**

Having looked at how we prepare our course, I now want to summarise the key aspects of classroom delivery. Of course, there is no substitute for practice but the next few pages highlight the key problem areas in CRM facilitation raised by delegates on CRM Instructor classes. I have assumed that all readers will be experienced classroom performers and are conversant with the issues of voice, posture, animation and so on; the bread and butter of classroom delivery. We often forget that the most sophisticated multimedia device in the classroom is the facilitator. As such, we need to remember that our performance will influence the impact of our message. However, I want to focus on 3 specific classroom skills: question technique, problem individuals and debriefing.

*Question Technique*

A facilitator is simply one who facilitates the learning of others (repetition – an event of instruction!). The facilitator assembles various appropriate resources and guides the exposure to, and interpretation of, those resources by the learner. The facilitator is, by virtue of their background and experience, also a resource. One of the most important tools available to the facilitator in this whole process is that of questioning. Sometimes called the Socratic method after a dead Greek, questions allow the facilitator to engage the student and to gain an impression of the student's level of understanding.

Questions can be categorised as open or closed. Open questions are typically those that start with what, why or when. The advantage of open questioning is that it encourages responses. Course members have to provide information or opinions in response to the question. On the other hand, closed questions are those that can be answered simply with a 'yes' or 'no'. Closed questions inhibit discussion. A closed question can, however, be used to force an opinion. Consider this interaction:

Are you saying, then, that in this situation we should consider doing ....?

This is a closed question that invites the respondent to confirm or deny their position on the issue in question. Having established the point, we can then explore opinions further through open questions. Inexperienced facilitators have a tendency to stick close to their planned script, which inhibits their use of questions. Because you are never sure what answer you will receive, using questions requires the facilitator to have a thorough knowledge of their subject so that they can deal with minor diversions off track. The use of questions in class throws up some additional requirements. First, and most important, is the need to listen. The facilitator must be sure that they have understood any question posed by the class. Quite often, a question carries an implicit meaning. Sometimes, the questioner may be looking for support for actions taken in some encounter out on the line. Sometimes, the class may be looking for support in some grievance with management. Especially when you are offered up as an outside expert, your opinions can be put to service in support of causes you could never anticipate!

The second requirement is the need to offer an adequate response. At the end of the day we get back to effective communication. If you didn't understand the question properly, then your response will be inadequate – so make sure you have understood the point! You may not know the answer, in which case simply say so. Quite often CRM debates are not about true/false situations but rather what would be effective in a specific situation. If we return to the role of the facilitator, the best answers are likely to come from the class. Rather than offer solutions, the facilitator should seek opinions from the class and then get the group to evaluate the various positions offered.

Question technique can often be criticised for a variety of reasons but 2 key crimes are asking rhetorical questions and leading questions. A rhetorical question is one that requires no answer. So, as we look at a picture of a smoking wreck, 'how good a landing was that?' seems a little redundant. Rhetorical questions do not require a response from the class. Like closed questions, they can be useful stylistic points but, if overused, the class can get confused as to when they are supposed to respond and when they should stay silent.

Leading questions are ones where the correct answer is implicit in the question. Quite often they begin with 'Do you think, then, that ...'. The problem with leading questions is that we never do find out what the respondent actually does think, only what they think about our suggestion. Flaws in question technique are common but if facilitators do not ask questions in the first place then we have slipped into 'lecturing', which is not the goal.

*Dealing with Problem Individuals*

In its short history CRM has already developed its own underworld – the boomerangs. These are students who reject the CRM message. Often, but by no means always, they are older Captains. I was running a CRM facilitators' workshop in a former division of Aeroflot when, one coffee break, an 'old and bold' Captain sidled up to me at the bus stop to the canteen. He was new to the group and we had not actually spoken before. Standing side by side, he took a long pull on his cigarette and said, out of the corner of his mouth, 'I guess you can't teach some old dogs new tricks'. Blowing the smoke out of his lungs he moved back into the crowd. On another occasion, during question time at the end of the final 2 days for the very last group of 104 pilots in a particular airline, a young First Officer asked me what he was supposed to do with a Captain who was worse now that he had completed my course than he was before I turned up.

So, we may encounter a problem in class but, to keep things in perspective, it will not be that often. The main types of performance you will meet are the difficult customers and the point scorers. The former simply challenge everything and the latter try to check you out by asking technical questions. In all cases, simply throw issues back at the individual by posing their point as question to them. For the difficult customers you simply ask them what their view is on the specific point they have raised and then ask the class for an alternative view. Do not get drawn into a debate. As for the point scorers, ask them to explain the theory they have encountered and ask them to explain how it links to the topic under discussion. In short, make the class work for you.

*Debriefing Skills*

As CRM develops from what was a classroom activity in the early days to what is now a more integrated training strategy, the need to debrief performance becomes ever more important. Of course, readers who are involved in line or simulator training will already have exercised their debriefing skills. For those ground instructors who want to run classroom LOFT-style courses, debriefing may be a new skill. In fact, debriefing is only an applied form of question technique.

The purpose of a debriefing is to promote learning through self-appraisal. A performance is analysed, interpreted in terms of success and failure and, finally, methods of remediation are identified. The important point is that the actors in the performance, not the outside observers, should do the analysis. There are various models of debriefing but we will consider one that works well and is straightforward. It comprises 3 stages. First, the individual or team who undertook the performance offer their opinion as to overall standard of achievement. It is important to get the players' view first. As soon as the assessor reveals their own view of the standard, the participants own interpretation becomes coloured. So, our opening questions might be 'How did that work for you? What were you happy with? What could you have done differently?' The assessor has offered no opinion of their own but, instead, has tried to ascertain the players' view of events. At this point, we need to prepare the players for what is to come and so the assessor identifies the best element of the performance and says 'I agree with you that … went well' or 'From my position, I thought … was especially well handled'. In fact, we are not too concerned with what went well. Usually we only have limited time available and we need to concentrate on the bits of the performance that were sub-standard. During the observed exercise we will have been noting down points for discussion. Before the debriefing we will have allocated some sort of priority to the key areas for improvement and will have highlighted those we want to cover in the debriefing. Having given some positive strokes, we now tackle the key area in need of improvement.

Again, we let the team do the talking. We might kick off with 'Let's look at … what were you trying to do?' The team will discuss their interpretation of the event and, through their explanation, we gain information about their thought processes. Our next question might be 'and how did that work out for you?'. By establishing a standard of performance we can then get the crew to review reasons for inefficient performance and possible alternatives. Finally we can move to direct questions related to why one course of action was chosen over another. Finally, we ask the crew what they plan to do if they encounter the same situation again. Through a structured sequence of questions, the players critique their performance, identify weaknesses and determine remedial action.

We said earlier that we should only attempt to analyse 2 or 3 key aspects of the performance. This is simply because any more than this would be to run the risk of lessons merging into some homogonous version of events in the minds of the trainees. We tackle the debrief points in decreasing importance, so, worst first and working through to least last. Finally, we come back to rewards. Finish the debrief by mentioning one or 2 positive points so that the players leave feeling positive and prepared to return for further training.

## Conclusion

In this chapter I have covered some of the practicalities of preparing training sessions and dealt with some of the core problems of delivering instruction. I am always surprised by how many people do not appreciate that facilitation is a skill that needs practice. I ran a facilitation course once where the delegates were surprised when I asked them to do practice sessions. Afterwards they admitted that, with hindsight, I was, perhaps, little naïve to expect to do a skills course without rehearsing the skills.

## What to do Next

All that now remains is for you to develop your course. Good luck!

## Reference

For information about visualisation of information see www.edwardtufte.co.uk.

### Table 5.3 UK CAA CRM Instructor Competencies

| **Plan and Design Training** |
| --- |
| A1.1 Identifies training requirements |
| A1.2 Identifies design and delivery resources |
| A1.3 Ensures facilities meet requirements |
| A1.4 Incorporates variety of media etc |
| A1.5 Involves other people in design |
| A1.6 Builds in methods of evaluating training effectiveness |
| A2.1 Identifies and selects CRM learning support material |
| A2.2 Ensures written and visual support materials are clear, accurate, practical and user-friendly |
| A2.3 Ensures activity and exercise materials are practical and realistic |
| A2.4 Prepares and presents durable support materials |
| A2.5 Promptly identifies and rectifies problems |

| **Deliver Training and Development** |
| --- |
| B1.1 Clarifies CRM in the context of overall training environment |
| B1.2 Makes links with technical training, SOPs etc |
| B1.3 Makes links with flight safety, customer service etc |
| B1.4 Makes links with similar training in other industries |
| B2.1 Establish CRM credentials |
| B2.2 Clarifies training objectives and methods |

| |
|---|
| B2.3 Ascertains and supports learners needs |
| B2.4 Continuously monitors and responds to changes in climate |
| B3.1 Clear and persuasive communicator |
| B3.2 Good manner and appearance |
| B3.3 Good presentation skills |
| B3.4 Organised, systematic lesson plans |
| B3.5 Clear, accurate presentation materials |
| B3.6 Actively clarifies understanding with learners |
| B4.1 Uses exercises and activities designed to maximise CRM training objectives |
| B4.2 Encourages trainees to get involved |
| B4.3 Clarifies roles, rules and expectations |
| B4.4 Gives timely feedback to trainees on outcomes and progress |
| B5.1 Recognises and responds to individual differences and problems |
| B6.1 Overtly supportive of CRM principles |
| B6.2 Motivating, patient, confident and assertive manner |
| B6.3 Encourages mutual support and teamwork among trainees |
| B6.4 Ensures learning opportunities for all trainees |
| B6.5 Includes teamwork, exercises and demonstrations |
| B6.6 Encourages sharing of individuals learning experiences |

| **Review Progress and Assess Achievement** |
|---|
| C1.1 Tracks trainees progress against formal benchmarks |
| C1.2 Conducts formative assessments based clearly on training objectives |
| C1.3 Reviews progress with trainees |
| C1.4 Sets new/additional learning objectives |
| C1.5 Keeps appropriate records |
| C2.1 Agrees and reviews plan for assessing performance |
| C2.2 Collects and judges performance evidence against criteria |
| C2.3 Collects and judges knowledge evidence |
| C2.4 Makes assessment decision and provides feedback |

| **Continuous Development** |
|---|
| D1.1 Tracks trainee performance against agreed criteria |
| D1.2 Tracks training session progress |
| D1.3 Elicits informal, ongoing feedback from trainees |
| D1.4 Elicits formal course evaluation from trainees |
| D2.1 Regularly reviews own performance, strengths and development needs |
| D2.2 Collects feedback about performance from others |
| D2.3 Attends CRM training conferences and workshops |
| D2.4 Keeps abreast of developments |
| D2.5 Maintains a written development record against a development plan |

# Part III
# Measuring Results

# Introduction

So far in this book, we have looked at what we mean by CRM and how to develop a course. The final question we need to consider is how can we tell if our efforts have had an effect? In this last section we will try to answer that question. In terms of facilitator skills, we will cover:

- A1.6 Builds in methods of evaluating training effectiveness

This section contains 3 chapters that look at methods of evaluating CRM training, methods of measuring CRM skills and, finally, the standardisation of instructors.

Before we start we need to consider what is meant by 'measurement'. Historically, facilitators have sought feedback on delegate satisfaction and suggestions for improving the course. Although important, 'happy sheets' are really measures of the surface features of training. If you accept that the whole point of CRM is to influence safety, then we should be looking at a change in behaviour that can be linked to a change in the safety status in the organisation. At the same time, one of the perennial problems for CRM facilitators has been the difficulty in convincing management of the need for training – despite it being a regulatory requirement in most countries. So, measuring the return on investment is another priority. These are the 2 areas we want to concentrate on throughout this final section. Ironically, such is the thorny nature of these problems that, although dealt with last in this book, developing your evaluation scheme will probably be the first thing you need to do when you think about rolling out – or continuing – a CRM programme in your company.

Chapter 6

# Measuring the Effectiveness of CRM Training

## Introduction

It has always struck me as something of a paradox that the introduction of the requirement to deliver CRM training has prompted considerable efforts to measure the benefits resulting from the training offered. One of the earliest conferences in the UK concentrated on making the justification for CRM in the eyes of management. Why is this paradoxical? In short, given that aviation is an industry that has invested heavily in technological training devices, such as simulators and computer-based training systems, it is rare that anyone ever asks questions about this even more significant investment. Research has been done into the effectiveness of simulators since Link produced his first trainer. We know that full motion and high fidelity daylight visual systems are not essential for skill development. A significant proportion of the CBT produced is mind-numbingly boring. But has the return on the investment in these training solutions been routinely measured? (For those of you who read Section 2 – that's an example of a rhetorical question). Given that CRM represents a relatively cheap investment in training, why, then, the frenzy over measuring the benefits? From the discussion so far in this book, the desire for measurable results is probably a reaction to the lack of any tangible improvement in safety directly arising from CRM training. In reality, the perceived benefits are probably reflected in the way airlines treat the topic. In one Scandinavian carrier I visited, cabin crew received 45 minutes of recurrent CRM every year. To make matters worse, it was the same presentation every year. In the US, many of the Part 135 Cargo operators manage to cover sufficient information in an hour-long initial CRM class to satisfy their requirements. In another US regional, the same day-long course had been presented every year for 11 years. If you upgraded to Captain or changed aircraft you had to sit through the course again. The record was one student who had to endure the course 4 times in 12 months.

Putting my cynicism back into its box, as a supposed training designer, I remain committed to the need to measure the effectiveness of any training intervention. In this chapter I want to outline what is involved in the process and to highlight the difficulties you will encounter. Unlike developing skills in a simulator, CRM is not thought of by some as having any clear outputs so, in some ways, asking what are the benefits of CRM training is akin to asking how long is a piece of string. Of course, in the previous 2 sections of this book we have tried to get a better understanding of what CRM would mean in terms of behavioural change. However, undaunted, let's get measuring. We will start by looking at ways of distinguishing between types of measurement before, then, moving on to look at how to measure.

## Kirkpatrick's Hierarchy of Effectiveness

The advent of a systematic approach to training design established evaluation as a stage in the design cycle. From the outset, good practice dictated that some attempt be made to ensure that training achieved its goals. Furthermore, and touching on points raised earlier, having data to support the effectiveness of your training system means that you can justify the investment to higher management, especially where there is competition for resources. A final reason for evaluating training is that it brings a degree of control over the training system. Cause and effect can now be linked and decisions taken about specific approaches to training. For example, we can answer questions about what level of technology to employ, how much time to spend in training and when to call staff in for training so that skills are sustained at the required level. With the increase in popularity of concepts such as 'Just-in-Time' training and 'Blended' training, questions of effectiveness become all the more relevant. The generally accepted framework for training evaluation is that of Kirkpatrick (1976). His 4-stage model asks questions about students' reactions to the course, the extent of their learning whilst on the course, their application of learning to the task and, finally, the benefits accruing to the organisation as a result of training.

*Level 1*

Level 1 evaluation, according to Kirkpatrick's scheme, uses student reactions to the course as a measure of success. Known as 'happy sheets', student feedback forms typically ask questions about how interesting the course was, how useful the information might be, what students thought of the facilities and so on. Level 1 evaluation is the most widespread form of evaluation and, in most cases, the only attempt made to measure the effectiveness of a training course. Student satisfaction is important. As a third-party training provider, I know that unhappy customers will not be back in a hurry! Happy students are more likely to pay attention in class. It would be wrong to trivialise the value of Level 1 evaluations but we need to sound a note of caution. Just because the course was enjoyable does not mean that the attendees will take the messages back to the workplace. I worked for one low-cost carrier in which the working regime was demanding. Students in the classroom were visibly fatigued and CRM was an opportunity to sit down and switch off. Course evaluations showed consistently high levels of satisfaction but I honestly think the students enjoyed the rest as much as, if not more that, they enjoyed my performance.

Quite often 'happy sheets' ask questions that students are not really in a position to answer. We regularly see students being asked what could be added or removed from the course or where was too much or too little time spent on a topic. To a degree, the content and duration of training sessions has been determined by our training design work. The students on the course may not be conversant with the regulations we need to satisfy or the logic behind the design decisions made when preparing the course. Happy sheets sometimes ask if the material covered on the course is relevant. However, having delivered enough CRM courses to new-hire pilots fresh out of training, I know that this audience is in no position to answer such a question.

Happy sheets are useful during the initial roll-out of a new course. As a rule of thumb, I like to run 3 iterations of a new course. During those 3 courses, I make no changes but I do gather feedback from happy sheets. Only then, once I have ironed out any wrinkles associated with the delivery of the course from my perspective, do I consider making changes based on the student feedback. I have worked with facilitators who, if a session does not work first time, will delete that event from the programme. We have all experienced an event in class that fails miserably. It is damaging to the ego and can shake your self-confidence. If the event fails 3 times in a row then there is a design problem. However, all events will fail once. It is as much a reflection of the willingness of the class to participate as it is a comment on the soundness of your lesson preparation.

Once the course is established and the fine-tuning is complete then I would advocate periodic sampling of future courses rather than a routine, mechanistic administration of happy sheets. After all, the forms need to be processed, which is an additional burden at the end of each course. That said, the decision may not be yours. It may be company policy to gather data on all training. They may have a standardised feedback form. The training programme may only be of a short duration. For example, many airlines try to get all their training out of the way during periods of low demand. Depending upon the size of the airline you may be required to complete the task in a few weeks and so the evaluation task is minimal. However, in the absence of company guidelines and certainly for programmes lasting more than a couple of months, I suggest that Level 1 evaluation be conducted every time there is a change of facilitator or some other change in course delivery. I also recommend random sampling throughout the duration of the course just to keep an eye on progress.

Level 1 evaluation is primarily concerned with customer satisfaction during the event and, as we saw earlier, satisfaction is not really an adequate justification for the expenditure. If we want to offer something more meaningful then we need to look at the transfer of training from the classroom to the workplace, typically what we measure in Levels 2-4. However, before we do that, it might be useful to consider what hard measures we can include in our happy sheets. Kirkpatrick has suggested that we include questions like:

> After this course I will be _____ per cent more effective at work

The delegate puts a value in the space according to how they perceive the course will shape their future behaviour. We can also get students to look backwards. We could ask:

> If I had been given this course earlier I would have been better able to solve problems at work – Strongly Agree/Agree/Not Sure/Disagree/Strongly Disagree

This is really a variation of the 'how useful' question designed to get students to reflect on their work situation. In both cases we can generate a value that we can revisit at some stage in the future. So, after 6 months, say, we can ask course graduates:

> I reckon I am more effective at my job as a result of having done that course – Strongly Agree/Agree/Not Sure/Disagree/Strongly Disagree

In the examples I have given we can convert the responses to numerical values for manipulation. The averaged values before and after should give us a feel for transfer. Although we have now started to stray into the next level of evaluation, these examples raise the issue of pre- and post training measures. If you have no standard of performance before training, then any attempt to measure training effectiveness will be difficult, if not impossible.

*Level 2 – Achievement in Training*

Evaluation at Level 2 attempts to measure what students have achieved during the training course. Typically, we would look at exam results as the benchmark of performance but, as yet, CRM remains a non-examined topic. In fact, some courses at pilot ab-initio level do contain tests and the JAA Human Performance and Limitations exam contains questions relating to what might call CRM. If I return to my lens/toolkit analogy, then the conceptual structures we plan to deliver as part of the 'lens' lend themselves to pencil and paper testing. However, factual recall is perhaps not the best measure we would hope for. It would be possible to let delegates evaluate a case study, using multiple choice objective questions (MCOQs) or free-form text as a means of capturing responses. This approach has been tried before but with poor results (Flin et al 2000).

If you recall the earlier discussion of 'boomerangs' – those course members who reject CRM – then clearly attitudes could be a target for change. It would be possible to administer some form of attitude measure before and after training and use that to indicate achievement. A number of studies have attempted to measure students' attitude to CRM as a domain in this way and many have demonstrated a positive shift at statistically-significant levels. So, could this be a useful measure at Level 2? The important question here is 'attitude to what?' Attitude to CRM as a subject is really little better than the 'did you find it interesting/useful' questions posed at Level 1. What we really need is some attitude that is linked to workplace performance and that can be shaped through CRM. In order to do this we need to identify exemplar behaviours that people exhibit in the workplace. We can distribute these behaviours along a scale from desirable to undesirable. We can then identify a set of beliefs held by people demonstrating these behaviours; beliefs usually made extant through verbal utterances or actions. With this information, we can begin to compile a set of questions that can tap into the attitudes we are hoping to change. In Chapter 4 I briefly discussed the problems associated with constructing attitudinal questionnaires. Developing test items that generate significant results – that is, can really discriminate between people on the basis of the desired quality – is an activity requiring specialist skills. However, we need to remember that all we are trying to do with evaluation at this level is get a crude measure of response to training. We are not developing psychometric tests. Care and attention in developing questions will, though, add to the value of the data we collect (see Foddy for an overview).

We should also remember that, at this level, we are only looking for an indication of achievement during training, not application to the workplace. So, despite being critical of measures of attitudes towards CRM as a subject, we could ask questions that consider the extent to which delegates can see how workplace safety can be enhanced through the study of CRM, whether they see value in broadening their understanding of the causes of process failure, if their likelihood of reporting near-miss events will increase after training. Of course, we are looking at indirect measures of achievement at this stage, with no guarantee that professed behaviour will be demonstrated in the workplace. That said, we might be able to demonstrate a correlation between achieving a significant change in attitude to, say, CRM and some positive change in workplace behaviour. In which case, we might be able to use an attitudinal measure as a reliable Level 2 evaluation index. We will return to this point later when we discuss constructing evaluation schemes. For now, lets look at training transfer.

*Higher Still – Changing Behaviours at Work*

Level 3 evaluations look at transfer to the place of work. For example, communication and decision-making are core CRM topics. In fact, communication, communication, communication could almost be the CRM training mantra. Quite often, when discussing the need for refresher training, managers have said to me 'they all had the communication lesson but they still can't communicate'. Decision-making is of interest simply because, with hindsight, we can easily spot the poor decisions crews made as the tragedy inexorably unfolded before them. So, here we have 2 specific skills that could be measured so that training-induced changes can be ascertained. If we are to measure change, though, as I said earlier we need some idea of the level of performance before training. Few, if any, training courses incorporate pre-training measures. For that simple reason alone, any discussion about the benefits of running CRM training can only be conjectural – and you will have noticed that the same argument must apply to level 2 measures.

Whereas at Level 2 we can administer the same measurement tool at the start and at the end of the course and see if there has been any change, with Level 3 measures it is a little more complicated. At Level 2 we are looking at behaviours we can measure while delegates are in training and, so, the process is easier to administer. At Level 3 our subjects have been released into the wild and there is a delay between training and measurement. Administration becomes more difficult and intervening variables will influence behaviour. Again, we will discuss this in more detail later.

At Level 3 we need to identify performance indicators linked to the behaviours to be changed. For example, we may be able to construct an indicator of communication effectiveness between flight deck and cabin crew. This could be a hard measure of frequency of contact between, say, the pilots and the cabin supervisor. It could be a softer measure, say, of perceived team atmosphere. To a large degree, establishing performance indicators requires lots of imagination and a good understanding of what is happening in the workplace. By clearly identifying either areas where people do not work effectively or stages in the work process that can be endangered through poor working practice, you can start to look for aspects of performance that can be captured. In many organisations, data is available relating to workplace injury and damage, absenteeism, delays, cost of service recovery to name but a few. If a piece of data can be demonstrated to have a linkage to the behaviour in question, then it might serve as a performance measure or help to create a more appropriate measure.

Although I have been suggesting that data from the workplace might help us establish the extent of training transfer, we need to remember that, at Level 3, we are concerned with exploring the extent to which the course content transfers to the workplace. In the section on Level 1 evaluations I did start to stray into Level 3 when I gave the example of the question we could pose to course graduates once they had returned to work. In a similar vein we can ask what tasks are now easier as a result of the training, where they have noticed an improvement in performance – or a drop in errors. During your pre-course analysis phase, the better you are able to operationalize CRM concepts – that is, turn them into things people actually do – the easier it will be to develop training transfer measures. The discussion of operational indicators leads neatly into the fourth category of evaluation: does it make any real difference in the greater scheme of things?

## The Payback

Performance changes at Level 4 – some evidence of benefit to the organisation – are the real goal of training. Here we are looking to demonstrate a multiplier effect. Training has not only produced staff capable of fulfilling their contractual requirements but, in addition, can add value through using initiative, solving problems, acquiring expertise, generating novel solutions and so on.

For example, a review of customer feedback asked respondents one question: 'would you recommend this company to a friend?' (Reichheld). The responses were scored on a 10-point scale. This is the sort of data commonly held by airlines. The study looked at historic data to see if there were any relationships between responses and company 3-year growth rates. The researchers calculated the percentage of responses scoring between 0-6 on the scale, whom they called the Detractors, and the percentage of respondents giving 9 or 10, labeled the Promoters. By subtracting one from the other a net promoter score was obtained. The financial figures for 10 major US airlines were correlated with the airline's Net Promoter Score. In 8 cases a net promoter score of <+20% was associated with zero or negative growth rates over 3 years. In the other 2 cases, positive growth rates were associated with scores of >+40%.

In the 2 airlines with positive growth rates, we can hypothesize that some component of the high customer satisfaction score reflects staff performance. It is also probably fair to say that all 10 airlines probably deployed some form of customer care training. Why did some work better than others? Of course, the mechanics of cause and effect in this case are obscure but it is evidence that some aspects of staff performance can be tracked through to the bottom line. I have already mentioned the disillusionment with CRM in terms of its lack of tangible results. From this example we can begin to see the difficulty of establishing cause and effect at Level 4 evaluation. There are often too many intervening variables to be certain which component of performance has most effect. We will return to this subject later.

In an ideal world, we would try to establish measures at each level within the framework. But we need to remember that evaluation is effortful and there is an internal return on investment calculation that needs to be made. At the end of the day, happy sheets could well be the cost/effective solution, given some of the difficulties we will now go on to discuss.

## What Evaluation Studies Exist?

Salas et al (2001) reviewed 54 published accounts of attempts to evaluate CRM training in aviation. They found that, whereas positive evidence was found in support of the first 3 levels, no study reported any signs of benefit to the organisation arising from CRM training. The studies reporting evaluation at level 2 tended to measure students' attitude to CRM rather than test scores. The UK CAA (2003) commissioned a review of CRM Evaluation strategies. This report looked at a sample of 48 papers, some of which were included in the earlier Salas et al study. This second study included some reports from other critical industries. Again, Level 4 evaluations were least frequently undertaken although some evidence of positive change was reported. Level 2 was the most frequently reported category. Here, some evidence was presented relating to recall of concepts but the majority of evidence used was attitudinal measures using a standard questionnaire. In a follow up study (Salas, et al, in press) evidence of attitudinal change was reported but the lack of systematic studies at multiple levels undermined the ability to demonstrate cause and effect. The authors highlight the lack of any clear evidence that CRM has had an impact on safety.

The 3 studies provide a useful source of models that can be adapted to suit your particular needs. However, as yet there is still no coherent framework for comparing studies. Every author uses a different experimental method, different measures of performance and results are measured using different units of output. We need to be wary of trying to make comparisons between different CRM interventions on the basis of evaluations conducted using widely differing methods. At Levels 1 and 2 it might be possible to develop a standard evaluation format to be used across different courses, thereby allowing us to compare different interventions, for example, a case study versus a video or a LOFT compared to a more structured lesson-based solution. At Levels 3 and 4 we will probably need to tailor evaluation to individual airlines, which will make measuring training transfer arising from specific types of intervention more problematical.

### Developing an Evaluation Strategy

By now 2 things should be clear: effective evaluation requires planning and evaluation activity must precede the delivery of training if it is to work. Hopefully, I have also demonstrated that we need to examine effects within the course and effects between the course and the real world. In terms of how we plan an evaluation, we can describe the ideal situation thus:

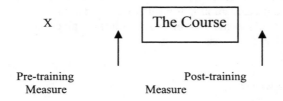

**Figure 6.1 Basic Evaluation Strategy**

If we also consider the need to demonstrate transfer then we can develop our model as shown in Figure 6.2:

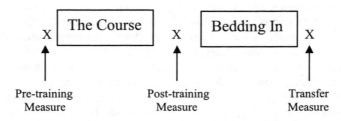

**Figure 6.2 Training Transfer Evaluation Strategy**

When we start to include the need for a measure of return on our investment (ROI), then the problem starts to become complicated. One model we can apply is shown in Figure 6.3. Here we can see that a comprehensive approach to evaluation requires the use of control groups, usually employees who will not receive training. We deliver the performance measure to both groups before training and then again at a set time after training. When we come to measure effects, we now need to look within groups and between groups in order to identify the importance of outcomes.

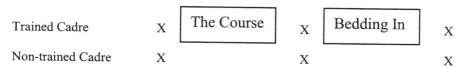

Trained Cadre    X    The Course    X    Bedding In    X

Non-trained Cadre    X      X      X

## Figure 6.3 Evaluation and Return on Investment Strategy

We do have other problems to consider at this stage. Some of the effects of training may take time to work through into practice and so, if we sample too soon, we may not get accurate data. In addition, all skills decay over time and so there is a risk that a sample taken too late will, similarly, give unrepresentative data. In some companies, CRM training has been running for years and so we need to identify what the pre-training measure is actually capturing. Probably, we will want to establish a point prior to rolling out a new CRM annual training package at which to capture data that can be used to measure the ROI for that years' programme. The issue of calculating ROI underscores the need to include our evaluation scheme in the project planning for the course.

**Raising the Evaluation Stakes**

With changes in the aviation market place, the need to maximise the ROI is paramount. I have worked with carriers that will delay implementing regulatory requirements on the grounds that to do so before their competitors places them at a commercial disadvantage. Kirkpatrick's framework – and its application in aviation – makes little reference to cash costs and returns. Given the new commercial imperative, the time has come to recast evaluation in terms of an airline's return on investment. In effect, we are now asking 2 questions:

Did the facilitators earn their keep?
Did the course earn its keep?

The first question addresses the competence of those responsible for delivering training. It considers the extent to which the task of facilitation was conducted effectively and it also looks at how facilitators promote the transfer of training. The second question cuts directly to the issue I highlighted at the start of this chapter – why are we spending money on CRM.

In recasting the evaluation question, we introduce a new problem. Talk of costs and benefits or outcomes maps onto the previous discussion easily enough but we also need to consider more directly the cause of any observed change in performance. If we cannot prove that the training we delivered resulted in the workplace changes then we have a problem.

For the rest of this chapter will look at the procedures we need to develop in order to evaluate effectiveness in cash terms. I start with the way in which we capture costs and then move on to the way we establish the return on investment. Then we will look at determining cause.

## Measuring our Investment in CRM

The cost associated with training fall into 3 main categories: one-off start-up costs, costs incurred each time a course is run and, finally, costs per trainee. Table 6.1 is a worksheet containing the most common training costs. It is not exhaustive but covers most items. Of course, the implication is that we can track and isolate costs and that all our costs are reflected in the calculation.

## Table 6.1 Training Cost Calculator

> A. One-off Costs
>> Course development (time) or purchase (price, fees)
>> Training needs analysis and research
>> Design and production of training plan
>> Design and production of training materials
>> Design and production of training evaluation plan
>> Per instructor (durables: videotape, software, overheads)
>> Projectors, computers, training aids, AV equipment
>> Fees to consultants or outside instructors
> B. Costs per Course Offering
>> Rentals (e.g., projectors, computers, training aids)
>> Telecommunications (e.g., Internet access)
>> Rental or allocated "fair share" use of classrooms
>> Trainers: travel, hotel, meals
>> Shipping of materials
>> Instructors, course administrator, program manager
>> Fees to consultants or outside instructors
>> Fees to evaluators
>> Support staff, e.g., audiovisual, administrative
> C. Costs per Course Delegate
>> Per participant (expendables: notebooks, tests)
>> Participants: travel, hotel, meals, childcare, parking
>> Participants (number of instructional hrs x average hourly rate)
>> Lost productivity (if applicable)

## *Hidden Subsidies in Training*

In reality, much CRM training depends upon the enthusiasm and commitment of the training team and many airlines' CRM programs contain hidden subsidies represented by the unpaid overtime worked by willing developers and facilitators. In addition, CRM practitioners share ideas and many airlines introduce CRM on the back of effort expended elsewhere in the aviation industry.

The Guide to Performance Standards underpinning the discussion in Part II of this book does, in fact, make mention of the need to recognize copyright issues. The ease with which materials can be copied does create an ethical situation, even if breach of copyright is such an everyday occurrence that most people probably do not even consider it a crime. I visited a large flag carrier airline together with a colleague who produced and distributed training videos. On a shelf in the office of our host were a dozen bootleg copies of my colleague's products not even hidden from view. I ran a workshop on interactive approaches to training during which the group participated in the Gate Game exercise illustrated in Chapter 4. At the end of the workshop, 2 managers from another large airline suddenly grabbed all the exercise materials from the table and literally ran to the door of the conference room. On another occasion I gave a demonstration for the CRM group in a small regional airline after which, I was later told, the Captain who led the team had tried to recreate from memory the exercise I had used. In reality, most of the content of a CRM class is public domain information assembled into a particular sequence for a specific purpose. Copyright is unclear in some circumstances but is crystal clear in others. Breaching copyright is easy to do but still represents theft. If I think back to my encounters with copyright pirates, all I can say is that the worse the safety record of the airline, the less they seem willing to invest in training. We can calculate the total cost of our training programme using the worksheet in Table 6.2. Calculating training costs is not isolated to CRM training, of course. In fact, we should apply the same discipline to all the courses we deliver.

We need to understand that making decisions about training is not necessarily a science, no matter how hard we try to reduce the process to a set of numbers. For example, some airlines feel that in-house trainers bring a better understanding of company issues even though they may be more expensive in real terms. Other airlines see that an outside facilitator will bring a breadth of experience from across the industry and can offer comparisons with other sectors of aviation. I have found that delegates often feel happier discussing internal issues with an outsider. Without a true understanding of the costs involved, though, we cannot guarantee to make the optimum choice between alternatives. Furthermore, without any idea of the costs involved, any discussion of the benefits of CRM becomes meaningless.

## Table 6.2 Training Programme Cost Calculator

A. Total of all one-off costs: _____
B. Total of all costs per course offering: _____
B1. Sum of B x number of times course is run: _____
C. Total of all costs incurred per course delegate: _____
C1. Sum of C x total trained: _____

Total Training Cost (sum of A, B1, and C1) _____

**Measuring the Return on Investment (ROI)**

Having identified how much it is costing us to deliver training, we can start to look at how to measure the benefits of that training. We need to be clear in our minds what we mean by a benefit. We may simply be looking to achieve compliance with regulations. Our controlling authority may have prescribed the content, duration and method of delivery of an acceptable CRM course. In which case the equation becomes one of how much we have spent in order to be compliant. We can then set the cost of compliance against the penalties for non-compliance and ask if the investment was worthwhile.

It may be the case that we have some control over the method of delivery of training. For example, we might want to compare the use of distance learning as oppose to a residential event, for example. We can now make comparisons between systems in order to see if we can remain compliant whilst using the lowest cost delivery system possible. There are several questions we can frame around the issue of cost of compliance but the thrust of this book is really about using CRM as an intervention strategy aimed at improved safety and efficiency in the workplace. Therefore, to limit our investigation of measuring the benefits accruing from training simply to compliance would be to undermine my thesis. We need to do more than this. I make the point to underline the importance of having a clear understanding of what you are evaluating before you start collecting data. Earlier, we set 2 tasks for evaluation: did the facilitator deliver and did the course deliver. It is to these questions we now turn.

*Did the Facilitator Deliver?*

I have said enough about the usefulness of 'happy sheets', but the need to consider the performance of the facilitator remains an important Level 1 evaluation topic. An ineffective facilitator in the classroom can undermine the message of CRM as well as destroy the preparedness of delegates to learn. In a traditional classroom setting there seem to be 3 characteristics of sound performance:

- The enthusiasm of the facilitator for the subject.
- The facilitator's knowledge of the subject.
- The facilitator's management of the classroom setting.

We can set these statements against a 5-point scale, thus:

The facilitator was knowledgeable.
Strongly Agree/Tend to agree/Neither agree nor disagree/Tend to disagree/Strongly disagree

The intervals convert into number scores between 5 (Agree) and 1 (Disagree) and these can be stored in a database. A score of 3 or less should trigger action by training management staff. Furthermore, we can ask for narrative comments for these grades of assessment that should then remove some of the randomness apparent in other facilitator grading schemes.

In our cost calculations we will have included a figure for the facilitator, whether this is an actual cash cost for an external supplier or an internal charge for an employee. Where we have a choice of facilitator, we can now start to measure the benefits accruing from the use of different categories of facilitator.   For example, an airline could chose to use Captains, First Officers, Flight Engineers, Cabin Supervisors, Cabin Crew or ground-based staff to deliver training (local regulations permitting, of course).

With an effective measure of facilitator performance and a cost for the use of each grade of facilitator, we can start to make decisions about who to employ based on something other than an emotional response. I still encounter joint flight deck/cabin training delivered by senior captains with no thought to the possible hidden message this situation conveys.   There are 2 questions we can set out to answer: how many facilitators should be used to present a course and what is the optimum source of facilitators?   The issue of how many is interesting.   The negative side of the equation is that it doubles the cost.   The positive aspects include an ability to role-play some of the skills of CRM during the running of the event, increased variety of delivery thus sustaining interest for the delegates and an ability to reinforce corporate messages (and I refer you back to my comment about joint CRM).   With input and output data I can start to address these issues.   I should add a caveat here.   At the risk of undermining my position on 'happy sheets', the question of 1 or 2 facilitators will also require some overall measure of course 'worth' to fully resolve the issue.   However, once the choice has been made, based on data, we can revert to collecting evaluation data than provides best evidence for management of training.   I discuss the management of facilitators more in Chapter 8.   Having considered the contribution of the facilitator to the process, we can now move on to look at the contribution the course makes to our overall company goals.

*Did the Course Deliver?*

To answer this second question we need to recast Kirkpatrick's typology. What we want to explore is the relationship between the changes brought about by the course and the impact those changes may have in the workplace. The nature of the relationship is shown in Figure 6.4.

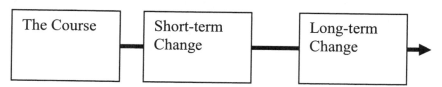

**Figure 6.4 The Benefit Stream**

The implication of this model is that the measure of achievement on the course, what we called Level 2 previously, and the measure of performance, Level 3 or 4, ought to be related in some way.  The ease with which this is done is in proportion to the thoroughness of our front-end training needs analysis.  If I have a clear set of performance objectives linked to a workplace goal then I can start to look for performance indicators.  However, whereas earlier, we looked at Level 2 evaluation as something occurring within the course, and even counselled against confusing Level 2 and Level 3, if we move to this revised paradigm, then we certainly need to consider relationships between performance indicators used during and after training.  In short, we want paired measures and a clear understanding of the operational output standard will make things easier.

The first step in the process is to look for the short-term measure.  I mentioned previously that training designers tend to think in terms of knowledge, skills and attitudes when establishing course goals and objectives.  We can use these categories to help us identify Level 2 measures and I briefly discussed some examples when I reviewed Kirkpatrick's typology.  However, what we now need to do is to consider objects within this schema that can be measured and that can be reliably established as intervening variables between the course and the real world.

**Table 6.3 Short-term Performance Measures**

| Domain | Performance Indicator |
|--------|----------------------|
| Knowledge | Working practices of other agencies<br>Informational needs<br>Probabilities of occurrences<br>Root cause of significant events |
| Skill | Of Communication<br>Of Decision-making<br>Of Conflict Resolution |
| Attitude | Towards risk<br>Towards safety<br>Towards own competence |

For example, imagine that I have identified improved working across organisational boundaries as a training goal.  My short term goal might be to increase the understanding of informational needs across the system in my course members.  I could develop an objective test based on specific forms of communication (unit x needs information y by time z in order to deliver their job function) and I can develop an attitudinal measure that taps into delegates' views about the benefits and significance of communication with other divisions within the organisation.  These indices will form my measure of short-term change.  I can then look for operational indicators linked to the concept.  For example, poor communication could lead to a task not being completed in a timely fashion, not being completed correctly or not being started at all.  This failure can be measured in terms of a delay to which we can attribute costs (compensation, missed connections, lost bags, processing of passenger complaints, requirement for additional crew if duty hours are exceeded, etc). These form the basis of my long-term performance measure. I have now created a mechanism for linking training to the bottom line. Tables 6.3 and 6.4 offer some examples of short-term and long-term measures.

In developing our measures we need to use some imagination. Measures can be either abstract, as with attitudes, or concrete, such as the cost of an avoidable workplace injury. What is important is that our measures are closely coupled such that variations in the short-term measure are tracked by the long-term measure. A failure to see this relationship indicates that our measures are most probably unrelated. As far as possible, I suggest you avoid accident data as a long-term measure!

## Table 6.4 Long-term Performance Measures

1. Time savings
1.1. Less supervision needed (hrs saved x supervisory $)
1.2. Better time management (hrs freed up x $)

2. Better productivity (quantity)
2.1. Less down time ($ value or reduced nonproductive time)

3. Improved quality of outputs
3.1. Fewer rejects (to scrap, lost sales, returns… $ value)
3.2. Valued added to output (bigger sales… $ value
3.3. Reduced accidents ($ value of savings on claims, etc.)
3.4. Reduced legal fees ($ value of savings on settlements)
3.5. Improved competitiveness (change in market share…$)

4. Better personnel performance
4.1. Less absenteeism / tardiness ($ saved)
4.2. Improved health ($ saved on medical and lost time)
4.3. Reduced grievances, claims, job actions ($ saved)

5. Increased Competitiveness
6. Enhanced Risk Management

## Picking through the Debris – What Happened?

Having collected our data, the first question we need to answer is did anything happen as a result of training? We do this by comparing our pre- and post-Training measures of performance and training. We need to ask 3 questions at this stage:

- Did change occur in the planned direction?
- Was the change significant?
- Was the change economically significant?

We are assuming that training will induce a change in a positive direction but we saw earlier that CRM can result in negative attitudes – the boomerang effect. Note that I have qualified the scope of the observed change: it must be significant and should show some economic benefit. These 2 tests are really attempts to answer the question 'so what?' If our training brings about no real change, then why bother. The economic test is, in a way, corroborating evidence in the event that the shift in the training measure is significant but the performance overall has little or no impact. For example, one of the selling points used to support the use of CBT in pilot type conversion ground school was that final test scores were often 3 or 4 percentage points higher on CBT than in traditional ground school. However, as all pilots invariably passed the test, irrespective of training method, this advantage was not economically significant.

We can address the first question by simple inspection of the data. So, if we expect a positive shift then the scores after training should be higher than those captured before training. If we want to look for a significant shift then we need to use statistical test, such as Mann-Whitney U Test and Pearson's Rank Co-efficient, to answer these questions. If the shift in scores is not significant then we need to ask about the effectiveness of the training given.

The final question asked if the change in performance was economically significant. This assumes that we have been able to convert our performance indicator into cash terms or some other unit with a commercial significance. Here, again, we need to be careful. For example, a decrease in passenger complaints could indicate an improvement in some aspect of service delivery and we could conceive of a commercial benefit accruing as a result. But an increase in passenger compliments could simply be the result of effective canvassing by staff.

## Cause and Effect?

The final step in our evaluation process requires us to investigate cause and effect. We need to be certain that any change is the result of our training intervention and not some other activity running parallel with training. I mentioned earlier some of the difficulties we might find with timing our evaluation intervention. Delegate behaviour could be the result of the bunching of interventions. For example, the measured performance could reflect the delegates' attitude to the company rather than their achievement in training. On one new-hire cabin crew course we were involved with, 7 out of 25 trainees withdrew from training at the half-way stage. For some young people, attending training could be their first venture away from home. Others have said to me that they never anticipated that the job involved dealing with medical emergencies, calming drunks or putting out fires. Elsewhere, a course for radar controllers regularly experienced 100% failure rates. Could this be bad initial selection of candidates or something wrong with the training system? Results such as these show the importance of being able to accurately determine cause and effect if we are to draw the correct conclusions.

We also need to ask if trainees are able to actually apply the skills developed in training or if there are organisational barriers that will prevent the adoption of changed workplace behaviour. For example, we did a follow-up audit of graduates from a supervisory management course and found that many of them, on return to the workplace, were told by their superiors 'I don't care what they told you on the course, here you do it my way'. We see parallels in the way First Officers exercise assertiveness once they have been through CRM. In some cases, assertiveness seems to result in FOs losing sight of the authority vested in the captain by the company. In other cases, Captain's feel uncomfortable with assertive FOs and believe that their status is being undermined. Organisational barriers need to be understood if ROI projects are to generate meaningful outputs.

For short duration training events we can be fairly certain that the conditions present at the start of training will remain constant. However, some training programmes may extend of 2 or 3 years. Over such a timescale, the workplace can witness many changes. Graduates from the final courses could be returning to a very different airline than the one encountered by graduates from the early courses. To make sense, then, of our investment in training we need to have a clear understanding of any extraneous influences that could have an impact of graduate performance in the workplace.

Robson et al provide a useful discussion of the statistical process we can apply to measuring effectiveness and Doucouliagos & Sgro discuss the techniques we can apply to determining causal relationships.


## Conclusion

In this chapter we have considered the mechanics and the implications of trying to measure the outcome of training. We have seen how the whole issue of CRM has been questioned on the grounds of payback and yet few airlines seriously measure the effectiveness of any of the training delivered on their behalf. To be fair, few organisations in any industry do. However, measurement brings control and, for that reason alone, should be a goal we aspire to even if achieving that goal presents difficulties.

Here are some alternative CRM training strategies:

- Do nothing
- Change airline management
- Recruit from another airline that already does CRM
- Raise entry level of workforce (insist that new-hires have done a recognised CRM course)

Although tongue-in-cheek, these are all valid alternatives to running in-house training. Without reliable methods of training evaluation, all we can say is that our regulator is unlikely to let us get away with the first option. As for which one is the best, we simply cannot say. For those managers unhappy with any investment in training you might want to discuss the merits of option 2. Again, in the absence of data, it has as much chance of success as any other solution.

**What to do Next**

This chapter has shown that evaluation is more complicated that designing the course in the first place; perhaps this chapter should have been the first in the book. If you haven't thought about how you are going to measure the effectiveness of the training you design, then start now! I have tried to give some idea of performance indicators you could look for but I have also stressed the need for creativity. Most important, though, is the need to think about the pre-training measure. Without that, the best evidence of effectiveness is lost.

**References**

Anon (2000), *A Consumers Guide to Return on Training Investment*, Forestry Continuing Studies Network.

CAA Paper 2002/05, published 23 June 2003, 'Methods used to Evaluate the Effectiveness of Flightcrew CRM in the UK Aviation Industry', United Kingdom Civil Aviation Authority.

Cascio, W. (2000), *Costing Human Resources*, South-Western College Publishing.

Doucouliagos, C. and Sgro, P. (2000), *Enterprise Return on a Training Investment*, National Centre for Vocational Education Research Ltd.

Flin, R., O'Connor, P., Mearns, K., Gordon, R. and Whitaker, S. (2000), *Factoring the Human into Safety: Translating Research into Practice*, HSE.

Foddy, W. (1993), *Constructing Questions for Interviews and Questionnaires*, Cambridge University Press, Cambridge.

Kirkpatrick, D.L. (1976), 'Evaluation of Training', in R.L. Craig (ed.), *Training and Development Handbook: A guide to human resources development*, McGraw-Hill, New York, NY.

MacLeod, N. (2001), *Training Design in Aviation*, Ashgate, Aldershot.

Reichheld, F.F. (2003), 'The One Number You Need To Grow', *Harvard Business Review*, December.

Robson, L.S., Shannon, H.S., Goldenhar, L.M. and Hale, A.R. (2001), *Guide to Evaluating the Effectiveness of Strategies for Preventing Work Injuries: How to Show Whether a Safety Intervention Really Works*, National Institute for Occupational Safety and Health.

Salas, E., Burke, C.S., Bowers, C.A. and Wilson, K.A., 'Team Training in the Skies: Does CRM Training Work?', *Human Factors*, Vol 43 (4), pp 641-674.

Salas, E., Wilson, K.A., Burke, C.S. and Wightman, D.C. (in press), 'Does CRM training work? An update, extension and some critical needs', *Human Factors*.

# Chapter 7

# Measuring CRM Skills

## Introduction

In Chapter 1 I commented on the shift in emphasis in CRM training from the input of information about safety and the causes of accidents to the assessment of output – the observation of behaviour. In the previous chapter we looked at the evaluation of training. One of the most obvious ways to measure the transfer of training is to go and watch personnel demonstrating the behaviour we have advocated and tried to develop. With that in mind, this chapter offers a more detailed discussion of Level 3 evaluation in a specific context. However, if we were to limit our examination of measurement to just that – the evaluation of training – we will miss an opportunity to add value to the organisation. Therefore, before we get to the heart of the matter I want to look at the reasons why we might want to measure CRM skills.

We can identify at least 4 reasons why measuring CRM skills might be a good idea. The first is because our regulatory authority might order us to do so. In the broader context of training design, we might want to check if our classroom training is having a positive effect – Level 2 and 3 evaluations. Third, in order to maintain or improve levels of service and safety we may want to develop personnel and, finally, we may want to make decisions about selection, retention and promotion of personnel. I want to explore these in more detail.

The regulatory interest stems from a legal requirement to ensure safe aviation. Regulators often have to encourage or coerce airlines and the other components of the aviation system to adopt safe working practices. Their principal change management tool is the mandatory requirement. Although the introduction of CRM within the FAA area of jurisdiction has been largely advisory, the inclusion of CRM markers within AQP (and in ATQP in Europe) is more a reflection of the philosophical starting point of AQP, which is an inclusive analysis of the piloting task. By definition, this includes the skills of CRM. Within Europe, the approach has been more directing. Once the requirement for CRM training had been issued, the evaluation of CRM skills seemed the logical next step.

Next, and as we saw in the previous chapter, measurement of outcomes is a task associated with accepted models of training design. We need performance indicators that will capture the effectiveness of classroom training and one such measure could be the behaviour of flight crew after training, measured against an agreed standard.

The percentage of aircraft accidents ascribed to human actions, as opposed to Acts of God or technical failures, has now passed into the CRM mantra and gives rise to our third reason for measuring. We have accepted that, on occasions, crews have clearly failed to demonstrate acceptable CRM skills. The logical conclusion is that we may therefore need to help crews to improve their skills. To achieve this goal, we need to be able to measure some notional current position on a continuum of performance and to track progress in relation to a declared standard.

Finally, airlines need to recruit new crew, occasionally make decisions about discarding existing personnel and, also, choose between crew for the purposes of appointment to management posts, selection for captaincy or retraining onto new aircraft types. An individual's possession of CRM skills may be of use in making that decision and, again, we need ways of measuring performance.

Our reasons for trying to measure performance, then, can be complex and, of course, the principles apply to all staff, from check-in to chief pilot. At present, most formal requirements to assess are limited to pilots and the rest of this chapter will focus on that task. However, everything we discuss transfers to other behavioural assessment situations. In this chapter we will look at the process of performance measurement, covering the methods we will need to apply and the problems we will encounter. There are 4 specific areas I want to address: sampling behaviour, the assessor, the measurement tool and the observational setting.

**Sampling Behaviour**

In any sampling regime, the more samples you take across a broader range of situations, the more valid your assessment of the criterion under investigation. Of course, 100% sampling is impracticable and too expensive. So we have to decide on the level of sampling we need to ensure reliability. An airline samples pilot technical skills during 6-monthly simulator checks and annual line checks. Accident reports frequently reveal weaknesses in pilot technical skills that seem to have been missed during check rides. So, 3 samples of behaviour a year represent a dubious level of reliability; many CRM assessment regimes call for a single annual snapshot of performance. I leave the reader to ponder the usefulness of such a sampling regime.

There are 2 approaches to sampling we can adopt: direct and indirect. The direct approach involves using methods that evaluate performance first-hand. The indirect approach uses secondary sources of data.

*Direct Methods*

The simplest method of capturing performance is to go and watch it for yourself. We will consider some of the difficulties associated with this later in this chapter. The major problem we face with this approach, however, is what is known as observer effect. The mere presence of a third party in an observer role will influence the behaviour of the observed crew. Simulator instructors are well aware of the fact that crews undergoing periodic checks are on their best behaviour and airlines that have compared flight data from pilots in both aircraft and simulator situations have found that crews do not fly the aircraft the same way they fly the simulator. For example, procedures are usually more strictly observed in the simulator. In a modified form of observation, known as Line Operations Safety Auditing (LOSA), the observers gain the trust of the observed crews by stressing the act that all data will be pooled to generate a fleet mean average of performance. The individual crew is guaranteed anonymity. Unfortunately, this would defeat the object of evaluating CRM.

If having an observer in place distorts the candidate's performance then one solution may be to remove the observer. We could get crews to report on one another or we could ask crew members to self-report. Given the problems that could arise if crews were to report on one another we can probably discount this solution. Self-reporting, on the other hand, is probably worth exploring further. Self-reports have been used extensively in psychological research and some assessment procedures used in personnel management, such as 360° feedback, use the method. Self-reporting would need some form of corroboration, a means of verifying the accuracy of the self-reports, but the approach does offer a data source.

If we take a broader view of the crew process, then third parties having contact with crewmembers may also be useful sources of data. Cabin Crew and Dispatchers, for example, could contribute to an assessment of an individual but, in this case, their observations could only relate to that sub-set of behaviours relevant to their interaction in the workplace. Direct reporting can provide the best evidence of performance but has some drawbacks. What of the indirect approach?

*Indirect Methods*

Several recent accident reports have drawn attention to that fact that some form of video record of the crew would have assisted in the subsequent investigation. Installing video cameras on the flight deck has been discussed in the aviation press and is being resisted by pilots. However, some means of recording a crew's performance for later analysis would offer an indirect means of skills assessment.

There are 3 types of data capture we could consider: video, as just discussed, the CVR and the Flight Data Recorder (FDR). In Chapter 4 we looked at some of the problems of using CVR transcripts. Devoid of context and stripped of non-verbal cues, a transcript does not represent best evidence. However, we can agree that the speech acts recorded do offer some measure of how the crew was performing. The routine analysis of recorded flight data to assess crew performance in CRM terms is still relatively unexplored. There are difficulties establishing combinations of aircraft parameters that could be influenced by crew effectiveness but that is not to say it is impossible. For example, if the time between selection of the first stage of flap and the recording of weight on wheels is less than a certain value then we might say that the crew had conducted a rushed approach and therefore their CRM skills were probably pushed to the limit.

These indirect methods offer fragmentary evidence. They require considerable effort to download and analyse. They are unlikely to be able to offer contemporaneous evidence; output from these systems will lag behind the assessment event. Nonetheless, they do offer another piece of the jigsaw puzzle that is crew performance.

*Sampling Across Time and Context*

Implicit in the discussion so far is the fact that crew performance is consistent across time and that we can capture performance and set it against a scale of effectiveness. From what we have said so far, we can probably accept that neither assumption would bear scrutiny. If performance is inconsistent and data capture methods are unreliable, then the need for multiple samples is reinforced. Work in the field of attitude and personality confirms that, although our behaviour has a degree of consistency over time, the predictive validity of a measure of behaviour is often unacceptably low.

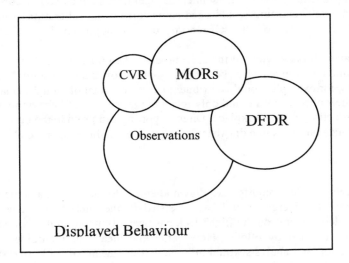

**Figure 7.1 Sampling Performance**

When developing our scheme we need to take into consideration the unit of measurement and the scope of the task being measured. By unit of measurement I mean do we look at the crew as a whole or do we assess each individual crewmember? If we place the focus on the crew as a unit of performance, then we can always ask if an individual would have performance any differently with a different crew pairing. If we take the individual as the unit of observation, then we need to consider the degree to which the individuals within the pairing exert influence over one another. Our question now is sorting out the extent to which the performance is representative of the individual being observed as opposed to an artefact resulting from the influence of the other players in the situation.

Finally, we need to consider the start point and end point of the observed performance. Many real-life CRM issues start in operations before the crew reach the aircraft but many simulator sessions miss the preparatory phase of the flight process. Equally, many runway incursions and taxiway incidents are a product of crew failures and this phase of flight, especially after landing, is absent from simulator check rides. Having discussed some of the issues around sampling behaviour, I will assume from now on that we will be using an observer to assess our candidates

## The Assessor

The observer we use to assess CRM are typically experienced pilots filling training or checking roles. It is assumed that years of operational experience will render these personnel fit for the task, often with minimal training. Unfortunately, even the most admirable of trainers and assessors is prone to unreliability on the basis of insensitivity and inaccuracy. By this I mean:

> Insensitivity: the observer fails to detect variations in observed performance
> Inaccuracy: the observer assigns wrong grade to the observed performance

This unreliability derives from 3 key areas. First, it is assumed that trainers and checkers have a thorough grasp of the domain in which they work. Second, it is assumed that assessors are familiar with the tools of their trade and, third, assessors are prone to observer bias. I will look at each of these in turn.

### Depth of Knowledge

Our observers need some depth of knowledge in the domain for 2 reasons: first, they need to be able to understand what is happening in front of them from a CRM perspective and, second, they need to be able to debrief the performance within a theoretical framework. The implication here is that the assessors have a sufficiently thorough understanding of CRM as a domain to be able to relate practice and theory in such a way that ineffective behaviour can be modified. In fact, this is often not the case. The observers often know little more about CRM than the crews being assessed. Quite often, they have attended the same classroom event as other crew and few, if any, have had any in-depth background training to prepare them for their role.

### Understanding of Standards

Standardised assessment can only work if all assessors thoroughly understand the framework being applied. Furthermore, assessors need to constantly refresh their understanding of the markers rather than depend upon their memory. Over time, memories can aggregate and distort remembered data. Audits of operational crews have revealed how a crew will apply a procedure they insist is in the company manual but cannot then produce the procedure – but they were sure it was there. We will see later that, in the absence of clear guidance, assessors apply their own standards to the candidate being observed. In much the same way, if the grade scale being used is ambiguous, assessors, again, apply their own standards. We try to overcome some of these issues through assessor training but reliable skills assessment requires observers to fully understand the tools of their trade.

### Observer Bias

Because observers are human, they are fallible. The functioning of memory processes, the design of the assessment procedure and the emotional and intellectual baggage observers' carry around with them all contribute to this unreliability. We can identify several sources of assessor fallibility and these are discussed next.

*Standards.* In the absence of clearly defined, readily understood and observable descriptions of behaviour, assessors will make up their own. Each assessor will have their own view of what constitutes an acceptable performance based on their past experience. They will have developed their mental picture of an ideal crewmember based on how they like to operate on the flight deck and incorporating the best, and eliminating the worst, traits of people they have flown with over the years. This is a natural process. The problem comes when all of our assessors are using their internalised models of good crew rather than the standard company model; our system lacks reliability. In this situation, the assessor formulates an opinion based on their own preferences and then simply writes the report to match their conclusions.

*Recency and Primacy.* When we try to recall events, quite often it is the opening sequences or the closing moments that are the most memorable. Everything in between tends to blur into an averaged version of events. In the same way, our recall of 4 hours in a simulator can be influenced by the first checklist we action or by the way we deal with the final emergency. The problem comes when the standard of the complete performance is seen in the light of the first or last behavioural elements we observe. Instead of treating each element in isolation and giving it a true value, we find that we develop an overall impression and assess components accordingly.

*Halo and Horns Effect.* This particular bias is based on the way we let prior knowledge of an individual colour our overall judgements. For example, in small airlines, I have spoken to crews who begin to picture a particular route and what the experience will be as soon as they see their roster. Based on their past experience of working with particular colleagues, they are already looking forward to a trip or planning how to avoid certain crew members. If our prior knowledge is positive, then we will tend to rate an individual more favourably – the Halo effect – whereas if we have bad memories of an individual we will see them negatively – the Horns effect. Assessors must recognise this flaw and consciously work to evaluate the observed behaviour of individuals strictly within the published framework.

*Leniency and Severity.* On old movie cliché is that of the good cop, bad cop. Most clichés have a germ of truth and the reality is that some assessors are more difficult to please than others. If 2 assessors viewing the same performance give different grades then our system is unreliable. If both assessors are the same side of the acceptable-unacceptable divide, then we can maybe live with that fact. However, we can have a situation where one assessor could pass an unacceptable performance and now we have a possible threat within the system. On the other hand, another assessor could fail an acceptable performance and then we have wasted training resources being applied to a problem that didn't exist. We will look more at grading scales later,

*Central Tendency.* Another problem we face with grading performance is that if assessors have difficulty in determining between levels of proficiency, the issue of sensitivity I mentioned earlier, then the most common outcome is for the performance to be graded as average. In some cases, over 90% of grades awarded were 'average'. If we go back to the reasons why we assess, if our motive is simply one of ensuring that our crews are performing to an acceptable minimum standard then this situation is not a bad thing provided our grading system is reliable. However, if we want to be able to discriminate between individuals for more specific reasons, then such a scheme would be useless. The use of an even number of intervals on a grade scale is often advocated as a way around central tendency.

*Scale Abuse.* No matter how many intervals exist on a scale, some assessors fundamentally believe that perfection is impossible and so the top grade can never be awarded. The converse is that some assessors do not believe that their organisation would ever employ someone so bad they would warrant the lowest grade. In effect, these assessors are grading on a reduce scale. We also see, in the absence of clear grade descriptors, assessors using half marks; they assign grades on the boundary between adjacent grades so that a 5 point scale now becomes a 9 point scale incorporating all the intermediate half grades.

**The Measurement Tool**

Having considered some characteristics of the people who will be doing the assessing, we now turn to the tool we will use for assessment. Our measurement tool is in 2 parts: the exercise and the performance capture tool. The exercise is the operational environment we will be using as the forum within which the crew 'acts'. This could be a LOFT scenario in a simulator or a normal line flight. The performance capture tool is the report form or grade sheet we use to record our assessments.

In Chapter 3 we discussed at length the concept of behavioural markers and I do not intend to recover that ground here. However, we do need to discuss the implications of using markers as measurement tools. The concept of a 'behavioural marker' is rooted in using observable behaviour to evaluate performance. By watching a person doing something we can formulate an opinion about their proficiency. We can make the assessment more significant by choosing tasks that are critical to the workplace performance being observed and isolating the skills associated with successful task completion. In Chapter 3 we looked at ways of identifying what we could observe and looked at some example markers.

*The Exercise*

Our aim, then, is to capture some significant aspects of performance against a set of criteria. The performance will comprise actions and words. These actions and words are the manifestation of internal processes. The route from intention to outcome may be short, direct and discernable in some cases, in other cases less so. That is fine, so far. We just need the setting in which to hunt for our behaviours. We have 2 options. We can set up an exercise in a simulator or sit on a flight deck and hope that 'behaviour' emerges or we can contrive a situation. LOFT, to a degree, establishes a set of representative critical tasks and, if designed properly, should have sufficient triggers to allow us to observe the behaviours we are interested in. The problem we face is that, with an average crew on an average day there is a risk that things might actually go well and we may not have anything observable to comment on. On the surface of it, this might seem a good thing. However, and we go back to sampling, if we are only making an observation on a single flight in a 12 month training cycle and that observation flight does not generate any useful data, then what can we really say about the usefulness of our monitoring system. If we take the other case where we construct a scenario in a simulator full of traps and mayhem, we are likely to generate vast amounts of 'CRM' but, again, how useful is data produced under such extreme and unrepresentative circumstances?

Our task is to develop a situation in which we can observe behavioural acts that have validity. By this we mean that the act observed is indisputably associated with the dimension being assessed. We also want to select acts that afford reliable assessment. By this we mean that if my candidate's performance represents a specific interval on my grade scale then, if all other factors are held constant, the next time I observe that candidate, the performance should attract the same rating. The observational setting will have an influence on the process and needs to be thought through. In many ways, the discussion so far should reinforce the need for multiple measures over time.

*The Marking Sheet*

The second element of the measurement tool was the marking sheet. In all probability this will be, in physical terms, no more than an set of boxes on an existing performance recording sheet associated with check rides and stored in a file. The important element is how the performance is classified or graded. Many airlines have a fairly arbitrary approach to grading any performance, be it CRM or the technical and procedural skills of aircraft operation. More often than not the guidance is simply 'we use a 5 point scale'. What the 5 points mean is a mystery.

I said earlier that one of the problems with observers is that they are inaccurate in assigning grades. I also said that 2 types of observer bias are attributable to problems with assigning a grade to a performance. The design of the grade scale is the key to reducing these components of observer unreliability.

There is no golden rule about how many intervals a scale must have. In some ways it goes back to your reason for assessing the in the first place. If your concern is simply to provide evidence of a minimum safe level of performance within the workforce, a 2-point scale (pass/fail) should suffice. However, if a more sensitive, graduated response is required, then meaningful intervals on the scale will have to be developed. However many intervals we opt for, there remain 2 problems to solve: how to describe the intervals and where to set the boundary between acceptable and unacceptable. I want to explore these problems by looking at an actual example drawn from an airline.

**Table 7.1 CRM Skills Assessed in Airline**

| Confidential Report | Line Check |
|---|---|
| Decision Making | Decision Making |
| Problem Solving | |
| Situational Awareness | Situational Awareness |
| Communication | Communication |
| Interpersonal Skill | |
| | Team Management |
| Command Ability/Potential | |

**Table 7.2 Grade Scales Used in Airline**

| Confidential Report | Line Check |
|---|---|
| 1 – Poor | Poor |
| 2 – Below Average | Minimum Expectation |
| 3 – Average | Standard |
| 4 – Above Average | |
| 5 – Outstanding | Outstanding |

The airline whose report forms I will be using raised an Annual Confidential Report (CR) and Line Proficiency Check (LPC) Report on each of its pilots. I have summarised the skills assessed on each report in Table 7.1. In Table 7.2 I show the intervals on the grade scale.

If we look at the instructions for completing the report forms we see some tacit recognition of the fact that the forms are to be used in different contexts; the LPC form comments on the requirement to note if manoeuvres are repeated during the check. That apart, the most obvious point is the differences in the skills to be assessed on each form. The skills are not elaborated upon or explained, nor is there any explanation of how 'Problem Solving' is to be assessed prior to the completion of the CR, given that there is no assessment of problem solving during an LPC. We could also ask how Team management differs from Command Ability. Of course, the reality of this situation is that report forms have been compiled by a variety of training managers over a period of time in the absence of any coherent model of effective performance. Forms like these are probably to be found in airlines the world over. Of greater interest to us is the construction of the grade scale. Here we see that there are 2 scales in use, one of 4 point and one of 5 points. That in itself is not a problem, rather it could simply be a rational management decision. What is interesting, however, is how the grades are described.

The LPC defines the 'minimum expectation' as a performance that meets the minimum requirement, something of a circular argument. Furthermore, this level of performance is less than that desired for effective crew operations. This would seem to be a contradiction in that if it is less than expected *for effective crew operations* then it cannot be the minimum 'expectation' – unless the minimum has been set below some notional safety margin. The comparable CR score of 'below average' accepts that the performance is acceptable but has flaws. The anomaly here is that the company is presented with a choice; the candidate can either work on those flaws *or* the company can keep a careful watch on progress. The implication being that a below-average grade need not require any action on the part of the assessed individual. Now, lest readers feel that the discussion so far is about semantics never forget that our observer will depend upon accurate descriptions to make our assessment system work. If any element of our measurement tool is ambiguous or can be open to interpretation by those using the system, then the scheme will be unreliable.

Moving up the scale, the LPC grade of 'standard' is described as that performance which would normally occur on a line flight. The CR score of 'average' describes a performance that is to a satisfactory company standard with no significant faults. We must assume that most pilots would be expected to be graded standard/average. However, the implication is that the assessor has a robust mental image of a 'normal line flight' against which to compare performance, which raises many questions about the risk of observer bias. We must also ask what is the difference between the faults that would be considered acceptable during an 'average' flight and the faults that would be considered 'flaws' for the award of a 'below-average' grade. Are we talking about quantity, severity, recognition and correction?

The CR score of above average has no equivalent in the LPC. It is a multi-standard score in that different expectations are set for inexperienced co-pilots and experienced captains. However, no guidance is given for when the standards should converge i.e. all pilots assessed against the same standard. Equally, the use of different standards according to experience could apply to all CR scores. Thus, a grade or 'poor' might be expected of some new FOs but clearly would be unacceptable for a captain.

The CR score of 'outstanding' is described as something that 'would not normally be used' as it describes a 'flawless performance'. So here we see a particular form of observer bias institutionalised. The LPC grade of outstanding lacks the caveat that it would not normally be used but describes a performance that few would ever expect to see: 'exceptional skill', 'model for teamwork', 'truly noteworthy'.

*Lessons So Far?*

When we set up a grade scale there is an assumption that each of the intervals on the scale represents a sort of mental bucket into which I can put the observed performance. What is more, we assume that the buckets are all the same size and are evenly distributed along the scale. The first thing we notice from this case study is that, in fact, intervals vary in size. We saw earlier that one form of observer bias is that of central tendency. Quite often, the design of the grade scale induces this bias because the 'average' bucket is huge in comparison with the other intervals on the scale. We have also seen that observers will lack sensitivity; they will not be able to distinguish between grades of performance. Again, the design of the grade scale can mitigate against the accurate capture of different levels of performance.

The case study illustrates the need for consistency across assessment where different performance capture tools are being used. It is not a requirement that all tools capture the same performance but where different tools sample the same dimension then consistency is required.

The main lesson to learn form the case study is the need for accurate, meaningful descriptors of grades of performance. This is not an easy task. The intervals need to be readily discernable and the distinctions between grades of performance need to be clear. Table 7.3 illustrates the grade scale developed for the NOTECHS scheme widely used in JAA countries. On the basis of the discussion so far, what do you think of it?

**Table 7.3 NOTECHS Grade Scale**

| Very Poor | Poor | Acceptable | Good | Very Good |
|-----------|------|------------|------|-----------|
| Observed behaviour directly endangers flight safety | Observed behaviour in other conditions could endanger flight safety | Observed behaviour does not endanger flight safety but needs improvement | Observed behaviour enhances flight safety | Observed behaviour optimally enhances flight safety and could serve as an example for other pilots |

First, the 'Poor' grade requires the assessor to be able to predict the future. How can anyone tell what might happen under other conditions? Perhaps different conditions would trigger a different performance from the crew. We can say what we see now and state whether or not the performance is a threat to safety. If we remove the predictive element then what is the difference between 'poor' and 'very poor'? Of course, we might argue that an experienced assessor could make a judgement about a crews' possible performance and that judgement might be reliable. The problem is, it remains no more than conjecture. It is not verifiable evidence.

Turning now to 'acceptable'. Here we see that the performance is safe but needs improvements. What we do not know is how much improvement might be required. What is the benchmark of performance? Are we talking about just CRM or are technical aspects included here? How do we distinguish between the 2? Research has shown that technical assessments can bleed across into CRM assessments.

The definition of a 'good' performance is that it enhances flight safety. The question is, in what way? Consider the gap between 'average' and 'good'. One recognises weaknesses but is 'good' an error-free performance? And how does performance 'enhance' flight safety exactly? If we try to envisage a normal crew on an average day, where would they sit between the 2 descriptions? Finally, what is the observable difference between enhancing flight safety and 'optimally' enhancing flight safety?

Here, again, we see the importance of choosing the words used to define the intervals on a scale. The descriptors we compile will influence the reliability of the assessments made. We need to make sure that the behaviours described are meaningful in the sense of being observable in an operational context. They intervals need to be discrete in the sense that an observer can assign a performance to an interval with no risk of confusion. We will look at how we can solve this problem later in this chapter.

*Establishing the Pass/Fail Divide*

There is a natural tendency to see the mid-point of any scale as being the divide between good and bad, acceptable and unacceptable. However, in terms of recording performance for either development or verification against a safety benchmark, this is not necessarily a meaningful way to look at a scale. Of course, we need an interval that represents a sound, acceptable category of performance. Why, though, do we want to record a better-than-acceptable performance? In all probability, we want to record above-average performers as part of some talent-spotting exercise for future selection of trainers, assessors or managers. Given that CRM assessment is not designed to be a personnel selection activity, we can realistically get by with a single category of 'above average'. What about the lower end of the scale? Here we are looking to distinguish between candidates for further training and candidates for disposal. Within the group who are deemed to need further training to attain the standard required, we might want to distinguish 2 subgroups; those that need minimal training resources and those that need some dedicated remediation. In both cases, it is implicit that the company feels that further investment in training is warranted. This approach gives us 4, or possibly 5, categories of performance we want to be able to identify and record and these are illustrated below:

Unacceptable | Needs Remediation | Standard | Noteworthy

Unacceptable | Requires | Remediated | Standard | Noteworthy
       Training   Post-flight

*Compiling a Grade Scale*

There are 2 approaches we could adopt in developing our scale. We can try to produce a single generic scale that can be applied to all dimensions being observed, which is the approach we have seen so far. Or we can develop specific scales for each dimension. The method we use to construct the scales in both cases is very much the same.

First, we need to get some experts do generate some statements of observed performance. The statements need to be brief descriptions of what they have seen for themselves during operations and training and must reflect differences in standard of performance. In the case of the generic scale, these statements will be summations of crew performance. In the dimension-specific case, the statements will be descriptions of specific behaviours relevant to the dimension. It is important that the descriptions are focussed, specific and, in the case of the dimension-specific approach, discrete in terms of the behaviours being described.

Once we have collected our samples of behaviour we write them on index cards. Then, enlisting the aid of our experts again, we mark out on a sheet of paper with a series of boxes arranged in a row. If we have decided on a 5 point scale then use 9 boxes, for a 4 point scale use 7 boxes. Number the boxes in sequence and mark Box 1 as 'low' and the highest-numbered box as 'high'. Ask each of our experts in turn to take the set of performance statements and to distribute them along the scale, placing each card in the box they consider appropriate for the grade of performance. Make a note of where statements are placed. Repeat this process with, say, 6 'experts'. Once the process is complete, use the statements in boxes 1, 3, 5, 7 and 9 to construct the descriptions of performance for our 5 point scale. The next step is to test the scale using standard assessors under normal conditions. The process I have described will turn a research scientist pale – but it will give better results than the methods used to constructs the scales we have looked at above. The important point is that we develop a scale that promotes standardised, accurate grading of performance across the assessor cadre.

## Capturing Performance – The Observational Setting

If we remember that our task is to capture observable data in a reliable manner, we need to consider some facets of the environment within which observations will be made. Before we do that we need to explore the concept of observable data in a little more detail. The output we are trying to capture is the result of an internal process. Output arises from motive – the participant acts as a result of some intention, albeit manifested in a deliberate act or through some subconscious learned response to a stimulus. Unfortunately, we cannot always identify the motive or intention from the observed output. We have another problem. Quite often, the extant behaviour is represented through speech acts. We need to deduce intentions from verbal utterances. To add to this, some behaviour (communication, team building, interaction style) exists in the space between individuals. Other behaviour (problem solving, decision making) exists, to a large degree, within the individual. When evaluating crew problem solving, for example, how do we allow for individual differences in stress tolerance, information processing (memory span, speed of processing, field dependency etc) and prior experience?

The performance being observed, then, is messy and obscure. In a training setting we could possibly intervene and seek clarification. This may not be possible, or even desirable, in an operational or a test setting. We could ask questions after the event but, then, we fall victim to problems of recall and reframing of events or rewriting of histories based on perceived interpretations – do I tell you what I was thinking or do I offer an explanation based on what I think you want to hear?

The environment in which observation takes place complicates matters further. In a simulator, the observer will also be controlling the training event and the need to attend to housekeeping duties, such as setting up malfunctions, can lead to CRM events being missed simply because the observer was looking the other way. The positioning of seats within the simulator or on the flight deck makes it difficult to see important cues such as facial expressions, eye contact etc. Given all the other demands placed upon, say, simulator instructors, we may have to limit the number of markers to be observed simply because the effort required to observe the full range will be excessive.

To compensate for some of these administrative, logistical and environmental issues, and to minimise the influence of observer bias, we need a framework for the process of observation. One that works well in other behavioural observation settings is given here:

O – Observe
R – Record
C – Classify
E – Evaluate

First, we observe the performance. This may sound obvious but for reasons discussed earlier the act of observation may difficult in itself. As an observer you will have an influence on the performance of those being observed; they will be on their best behaviour. The next step is to find a method of effectively recording what we see. Assessors will need to develop a mechanism for capturing key points of evidence. Note taking is, itself, intrusive and contributes to observer effect. Note taking is also a distraction from the act of observation. Each assessor will develop their own process and many experienced line trainers will probably have developed a technique already for keeping track of the procedural and handling aspects of performance. Observation and Recording happen during the observation event. Once back on the ground we enter the second stage of the process.

I discussed probable sources of observer bias earlier. One way to reduce the influence of the various biases is to separate data capture from data analysis. We do this by delaying the judgemental process as long as possible. Once on the ground, we review our notes and elaborate on our scribblings whilst information is still relatively fresh in our minds. The then classify our notes, placing the observations into the relevant behavioural category. With all the evidence to hand, finally, we evaluate and assign the appropriate grade to the performance.

### Training Assessors

Behavioural assessment cannot work unless assessors are trained first. Any airline that chooses to introduce CRM assessment without first training its cadre of line training captains or simulator instructors is simply ensuring that the process will be conducted in a haphazard and unreliable fashion. Assessor training has 4 components. First, assessors need to have time to fully understand the process of assessment and the tools they will use to capture and grade performance. Second, they must be made aware of all of the factors that can bias the observer or skew the observed performance.

With this background information, assessors then need to rehearse the process. This is done by using exemplars of behaviour, usually a video of a performance, and getting trainees to assess what they see. There are 2 stages to this process. First, assessors need to practice observing performance and then they need to practice the skill of assigning a grade to performance. In fact, we are not simple developing a skill at this point, we are also standardising. We need to ask trainees to grade a performance and then share their grade with their peers. Where there are discrepancies between assessors, these need to be resolved within the class through discussion. Individuals need to explain their grading through the use of observed evidence.

The final part of assessor training is rehearsing the skill of debriefing performance. One of the main purposes of assessing is to develop the individual. Therefore, assessors need to be able to conduct a developmental debrief that allows the individuals to accurately analyse their CRM performance, to identify shortfalls in performance and to establish more effective modes of behaviour for future use. However, if behavioural modification is to be effective, the individual trainee needs to conduct this process guided by the assessor, who now becomes a facilitator of change.

### Special Situations

The requirement to assess CRM is spreading across all sectors of aviation. Unfortunately, some niche sectors present different challenges for assessors. I want to look at 2 specific examples: single pilot operations and smaller operators.

*Single Pilot Aircraft*

Few operators of single pilot aircraft have access to simulators and so initial and recurrent training is often conducted on the aircraft. Given that this type of operation employs procedures designed for single pilots, the mere fact that 2 pilots are now involved will have an impact on performance. The first issue, then, is the fact that observer effect will be amplified in these situations. The next [problem is the extent to which behavioural markers developed in a multi-crew environment map onto the single pilot situation. The skills of piloting, both in terms of physical control of systems and airframes and in terms of the cognitive control of the flight process, are fairly constant across different crew configurations as I tried to demonstrate in Chapter 3 when we discussed Cessna Caravan operations. However, how the internal processes are manifested is even more obscure in the single pilot community than in the multi-crew situation.

Assessing CRM in this type of operation, then, will be difficult if we want to meet the criteria of being valid and reliable. One solution might be to use self-reporting as a means of collecting data (Burdekin). This method has been successfully used for single-seat military pilots.

*Small Companies*

Small companies with, perhaps, one or 2 aircraft have an added problem. The number of pilots will, similarly, be small and the pairing of line crew with training captains is routine. When a pilot returns for recurrent training, the assessment undertaken by the check pilot often incorporates aspects of the performances witnessed during the previous normal duty flights. A check ride is not simply a snapshot in time, it is a summation of performance over a prolonged period. The first problem we encounter, then, is that of prior knowledge being incorporated into assessments.

Smaller companies often employ smaller aircraft. Where an aircraft lacks a jump seat, the observer is usually also part of the crew. In this case, we not only have issues of observer effect but, because the observer is a crew member, the performance of the observer within the crew process could, in fact, initiate a CRM problem. Some airlines have tried to compensate for this by having the observer seated in the first row of seats and listening on a headset. Although this is practicable for smaller aircraft with no flight deck door, we have simply made the observational setting even more constraining.

**Conclusion**

This chapter has attempted to map out the difficulties you will encounter when you start to measure CRM skills. The assessment of CRM skills is an evidence-based approach. This means that an observer collects examples of behaviour against a scale. Such an approach places certain responsibilities on assessors. We need to collect reliable scores, or grades, and we need a coherent narrative. Personal feelings and beliefs play no part in an evidence-based system. The acid test of such a system is to imagine having to defend an assessment of an individual before a tribunal. For every judgement made there will need to be reliable evidence. That evidence will need to be linked to an argument that proves that the behaviour being measured is an essential requirement for safe or efficient aircraft operation.

In this chapter I have provided an overview of all the issues an airline's CRM assessment system needs to address. The problems associated with the measurement of soft skills are considerable. One golden rule to apply in trying to develop a workable system is – keep it simple!

## What to do Next

The process of measuring CRM in the workplace is, initially, a change management project and requires planning. As part of your CRM development process, you need to think through how you intend to roll out the assessment scheme.

What vehicles exist within your organisation for communicating change? What stakeholders need to be involved form the outset? What long-term problems can you envisage?

Time to draw up a list!

## Reference

Burdekin, S. (2003), 'Mission Operations Safety Audits (MOSA)', *Aviation Safety Spotlight* 4, pp 23-29.

# Chapter 8

# Administration of the Process

## Introduction

In this final chapter I want to tidy up some loose ends and, therefore, this will be something of a compendium. I want to discuss some of the background issues that will have an impact on the effectiveness of the CRM training and evaluation system we develop. The 2 main themes of this chapter are change management and sustaining the system once it is in place.

## Managing Change

Although CRM is not new to many countries, there are still some airlines who have yet to implement or who have a fairly basic approach to training in place. However, CRM skills assessment is still very new and, as such, the concept needs to be sold to the workforce. This requires pilots (initially, although all safety critical staff eventually) to know the why and how of the process.

**Table 8.1 Change Management Activities**

| | |
|---|---|
| Clearing the way | Clarify the aims of the scheme<br>Establish the processes involved |
| Breaking down Barriers | Allay fears<br>Identify the checks and balances in the scheme |
| Keeping Control | Who has authority?<br>Involve all decision-makers from the outset. |
| Taking Decisions | Document actions and decisions.<br>Communicate with Stakeholders |
| Implementing Change | Prototype everything<br>Seek feedback |
| Learning for the Future | Review progress<br>Record lessons learnt |

Any change management project involves several processes running in parallel with the actual change being implemented. These are summarised in Table 8.1. I have also included an example of a letter for possible distribution to staff affected by a skills assessment programme (page 170) and an example of the chapter you need to be thinking about for inclusion in your company training manuals (page 167-169). Although these 2 documents are framed around pilot training, they can be modified to suit any work group.

## Sustaining the Integrity of the System

By assigning a numerical value on a grade scale we are not simply collecting data about the candidate being assessed, we are also collecting data about the assessor. We can use that information to draw conclusions about the reliability of the cadre of assessors we are employing. By using statistical techniques (Pearson correlation) we can look at 2 key properties of our assessors: the extent to which their assessments co-vary and the extent to which they cluster around the standard we have set.

The first characteristic is called the Inter-rater reliability (IRR). We compare the scores of our group of assessors to see if they all fall within the same bounds. Because we are looking at a set of scores for each rater, where an individual is statistically different from the group as a whole, then we need to ask if the issue is the candidate being assessed or the rater themselves.

Referent-rater reliability (RRR) looks at the degree to which the raters' scores agree with the scores assigned by experts to a specific performance. In order to calculate this statistic we need agreed referents; examples of performance on video that have been rated by experts. Without this tool, we cannot consider using the referent-rater approach.

Theoretically, the 2 statistics ought to correlate one with another but, needless to say, there are sources of unreliability. In chapter 7 we looked at observer bias. It could be that the assessor group as a whole is biased in a specific area and so, although they show highly correlated scores for a specific parameter, we might find that the group as a whole is skewed and so the score is unreliable. We might find that the descriptions of the intervals on the scale is ambiguous and so the referent-rater statistic is unreliable.

Of the 2 approaches RRR is better but does assume we have been able to develop the benchmark performances before we start training. Using the RRR, we can better standardise assessors before we send them out into the world and it also allows us a more controlled vehicle for recalibration, which I discuss next. But, before that, we need to consider a final problem. Because if the traditional approach to training that has concentrated on the technical skills of aircraft piloting, we are already seeing a contamination of CRM grades. Beaubien et al compared scores given during simulator exercises. Each sortie was divided into stages and a score given for technical skills and for CRM skills during each phase. Results were found to cluster for the event being observed, not the skills being assessed. The exercise was generating unreliable data. The implication of this report is that we should also compare CRM scores with technical skill scores if we want to have confidence in our system.

## Need for Recalibration

In the previous chapter I looked at various forms of observer bias. On feature of any behavioural assessment scheme is that, over time, the attitudes of assessors towards their subjects harden. This can be tracked through a progressive lowering over time of the average grade scores awarded. I call this the 'gatekeeper syndrome'; assessors become evermore difficult to please, seeing it as their duty to keep the below-average world at bay. Of course, the quality of the candidates under review has not changed at all; the assessor has.

To overcome this problem we need to plan on an annual recalibration of assessors. We can do this by using referents (see Baker and Dismukes) or by using videos of crew performance taken, say, from simulator exercises. Groups of evaluators review and grade the performance and then share their observations in much the same way as they were initially trained and standardised.

In some situations, periodic rewording of the grade scales has been found to improve the scores awarded. It seems that the changes are sufficient to force assessors to pay renewed attention to their task. A failure to recalibrate will lead to your assessment scheme becoming unreliable.

## Remedial Training

I have talked at length of the process of assessment but have said very little about what to do if a candidate performs below the desired standard. I have also mentioned that group of personnel who reject the message of CRM. Clearly, an airline needs a vehicle for dealing with that small segment of the population who do not display the behaviours we expect. You will need a remedial training process.

Remedial soft skills training present a number of problems. We are concerned that an individual does not act in a desired manner. The problem is, in the first place, to identify the reasons why. It could be that the individual chooses not to act in a particular fashion or it may be because they cannot demonstrate the desired behaviours.

If an individual chooses not to employ the desired skills then this is probably a reflection of that individual's attitude. It is also, probably, an artefact of the individual's degree of self-awareness. The person fails to see the impact of their behaviour on others or is insensitive to the factors that influence their own performance. Quite often, these individuals are older, or more senior, in the organisation and do not see the need for them to change; their status does not require them to change. Where and individual cannot deliver the desired performance then we are clearly in the area of individual skill development.

In either case, the only effective method of remedial training will be individual counselling. The candidate needs to be given the opportunity to assess their current standard, identify weaknesses and then rehearse alternative modes of behaviour. Video from simulator exercises would be the starting point for any discussion and peer comments might also form a valuable source of information.

Remedial instruction for CRM is still in its infancy. Models can be drawn from other industries and from counselling generally. The important point is that an airline will need a mechanism to deal with substandard performance and assigning individuals to additional classes will have little value in terms of bringing about change.

## Continuous Professional Development

Like any field of endeavour, CRM does not stand still. New events that better illustrate CRM principles are happening every day. Research that elaborates underlying principles if not quite pouring out of university departments is still, nonetheless, emerging. Other sectors of industry offer parallels and insights that throw light on aviation. However, keeping up to date takes time. Conferences that address the specifics of CRM are surprisingly uncommon and the 2 key journals were listed earlier.

Few countries have introduced any form of accreditation specifically within the CRM field (the UK being the only one so far at time of writing). However, several countries have systems of professional training and accreditation controlled by national bodies so mechanisms for training to recognised standards are widely available. The use of professional accreditation schemes is one way to continue the development of facilitators.

Professional development can easily consume 2 days a month.

## Conclusion

We have now seen all the components of a CRM training system. In chapter 1 I suggested that our goal was to develop efficient workers who, as a result, would, in all probability, be safe workers. I have said almost nothing about training for workplace efficiency, instead concentrating on the 'safety generation' behaviours. That said, the theme of the whole book has been the need to see safety and production as 2 sides of the same coin. Most of the principles addressed in the book apply equally to technical skills training and, as I have said, to all work groups engaged in aircraft generation, dispatch and control. The task of building safe systems in aviation is complex and I hope this book has gone some way to clearing a path through the process.

## References

Baker, D. and Dismukes, R.K., 'A Gold Standards Approach to Training Instructors to Evaluate Crew Performance', NASA Technical Memorandum. 212809, Moffett Field, CA: NASA Ames Research Centre.

Beaubien, J.M., Baker, D.P. and Salvaggio, A.N. (2004), 'Improving the Construct Validity of Operational Simulation Ratings', *International Journal of Aviation Psychology*, vol. 14 (1) pp 1-17.

Johnson, P.J. and Goldsmith, T.E. (1998), 'The Importance of Quality data in Evaluating Aircrew Performance', http://www.faa.gov/avr/afs/ratterel.pdf

## Example Staff Letter

### Introduction

Following the introduction of training in Crew Resource Management (CRM), the next step is to evaluate the effectiveness of crew performance under line conditions. This serves 3 main purposes: it meets our obligations under JAA requirements; it affords the company the opportunity to verify that we operate to the highest safety standards; it allows us to validate the CRM ground instruction provided. With effect from {date} we intend to introduce formal assessment of CRM skills. This letter describes the process we will adopt.

### The Assessment Process

In order to assess CRM performance it is important to establish in fairly precise terms the areas of behaviour under scrutiny. To this end we will initially adopt the NOTECHS framework developed with UK CAA and JAA support. The framework will be described in an amendment to the {manual}. During OPCs and Line Checks {to be confirmed} pilot performance against the framework will be assessed and debriefed as required after the check. A behavioural marker framework provides a common baseline against which to discuss performance. All future CRM ground instruction will use the framework to direct the development of training material.

Because human behaviour is not clear-cut, nor readily compartmentalized, we have put some safeguards in place to make sure the process is conducted in a reliable manner. First, the marker framework is a public document accessible to all. The assessors have been trained in the use of the markers and will undergo periodic retraining to ensure that the assessment process remains true to its intentions. Finally, the reports raised by assessors will be quality-controlled to ensure standardization.

### The Implications of Assessment

The spirit of the regulations as explained by the CAA is that no pilot will fail a license or type rating revalidation due to poor CRM. It is anticipated that poor performance will be reflected primarily in a technical failing, possibly aggravated by deficient CRM skills. However, it is possible that an individual will achieve an unacceptable grade on 1 or more of the CRM markers. In the first instance, it will be the responsibility of the Training Pilot to resolve any issues of poor performance during the post-assessment debrief.

In extreme cases it may be necessary to offer remedial training in specific areas and nominated company pilots will be trained to offer such support as is necessary.

In the case of a dispute over an assessment, an appeal process to an independent third-party will be provided. Full details of the remedial and appeal processes will be contained in {manual}

### Conclusion

Aviation has traditionally assessed the hard, technical skills of piloting. The increasing recognition of the importance of the soft, CRM skills has forced us to take a broader view of what constitutes an 'expert' pilot. The system we plan to introduce addresses our immediate regulatory requirement in a manner that is transparent and offers benefits to all.

# Template Company Manual Insert

{company name}
Operations Manual (Aeroplanes/Helicopters)
Appendix to Part D

## Crew Resource Management Training

Reference:

JAR-OPS Subpart N; JAR-OPS 1.943.
B. JAR-OPS Subpart N; JAR-OPS 1.965.
C. CAP737

## Management

Captain {name} is nominated as the Company CRM Manager.

[Norman MacLeod, of Kitty hawk Training Technology, UK CRMIE 10035 C, is approved to deliver CRM training on behalf of {company name}{**required if third party training provider used**}].

*CRM Line Training*

Captain **{names}** are nominated as CRMI (Line) for the purposes of assessing CRM skills.

*CRM Ground Training*

**{names}** are nominated as CRMI (ground) for the purposes of CRM Initial, Recurrent and Captain's Upgrade ground instruction.

*All CRMIs require:*

Basic instructional techniques training (or core course).
To have attended a CRM course
CRMI (line) need to be trained to evaluate and debrief CRM skills.
CRMI (ground) need to be trained to deliver CRM in the classroom.

## Syllabus for Initial Issue of CRM Certificate

The Initial CRM training is to be conducted in accordance with Reference A and should include:

Human error and reliability, error chain, error prevention and detection

Company safety culture, SOPs, Organisational factors
Stress, stress management, fatigue and vigilance
Decision making
Workload management
Communication and co-ordination inside and outside the cockpit
Information acquisition and processing, situation awareness
Leadership and team behaviour' synergy

## Recurrent CRM Training

The Recurrent CRM training cycle (Ref. B) will refresh the above subjects over a 3-year cycle in accordance with the following schedule **{example only}**:

Year 1
Company safety culture, SOPs, Organisational factors
Stress, stress management, fatigue and vigilance
Decision making
Workload management

Year 2
Safety Culture
Human error and reliability, error chain, error prevention and detection
Communication and co-ordination inside and outside the cockpit

Year 3
Safety Culture
Advanced Communication
Information acquisition and processing, situation awareness
Leadership and team behaviour' synergy

## Assessment of CRM Skills

*Introduction*

The evaluation of behaviour in the workplace is conducted, first, to improve levels of safety and, second, to develop the highest levels of performance within operational crews. Any system of evaluation requires a structured approach to the observation, recording and interpretation of crew behaviour and the assessment of performance against a standard. This section describes how flight crew CRM skills will be assessed in {airline}.

*Administration*

Assessment of CRM skills will be conducted during Line Checks and Operator Proficiency Checks **{as required}**. Assessments will be conducted by qualified personnel in accordance with the framework of behavioural markers described in {**para number**}. The performance of each crew member will be observed and a record kept on individual training files.

{**Forms**
    **Example**
    **Instructions for completion**
**Instructions for disposal**
**What is going to be different from standard LC/OPC paperwork?**}

Where necessary, CRM skills will be debriefed after the checkride. If, in the view of the assessor, an individual pilot requires additional development to achieve the required CRM performance standard, training will be provided under the remedial scheme described in paragraph {}. In the event of a dispute, an independent appeals process has been established and is described in paragraph {}.

The {**company position**} will review CRM assessment reports every {**6 months**} to verify standardisation of assessment and to identify trends for referral to the CRM Manager.

    {**Data Collection**
        **Forms**
        **Administration**
        **Disposal**
        **Review frequency**}

*The Marker System*

The behavioural marker framework to be used is described in Appendix 1 {**NOTECHS as default**}. The framework provides a common terminology for discussing CRM performance as well as a structured system for capturing CRM performance. The company reserves the right to adapt or develop the marker framework according to changes in operational circumstances.

CRM skills will be graded using the scale described in Appendix 1.

## Training and Standardisation

Selection. All company training and checking staff, together with representatives of third-party training providers who fulfill these roles on behalf of the company, will be required to complete an appropriate course of training.

Training. The syllabus of training for CRM Assessors is described in Appendix 2.

Standardisation. Because behavioural observation is not an exact science and because assessor skills degrade over time, standardization of assessment is vital if the scheme is to achieve its goals. The primary vehicle for achieving standardization is the initial training course. The standardization process is supported through 6-monthly reviews of assigned grades. Finally, all CRM assessors will undergo annual recalibration.

## Remedial Training

In the event of a crew member being awarded a CRM grade of below average on one or more markers, the check report will be forwarded to the CRM manager. Having discussed the performance with the individual concerned, a course of action will be agreed between the CRM Manager and the individual.

If an individual fails to respond to training, formal warning procedures will be initiated {**this needs elaborating and checking against employment law**}

The stated intention of the CRM assessment scheme is to develop staff. Therefore, if the individual on whom an adverse report has been raised wishes to appeal against the report and independent procedure has been established. An independent third party with experience in CRM assessment and training will review the available evidence and act as an arbitrator.

## CRM Training Audit

The assessment of CRM skills offers the opportunity to evaluate the quality of CRM training within the airline. Therefore, the CRM Manager will undertake periodic reviews of CRM assessment grades and narrative comments in order to align CRM ground training with operational requirements. An annual CRM training review will be conducted as part of the planning for the subsequent CRM refresher programme.

# Index

advanced qualification program, 4, 8, 143

aeronautical decision-making, 5

alternative training and qualification programme, 8, 143

AQP. see advanced qualification program

ATQP. see alternative training and qualification programme

barrier analysis, 13, 14

cabin crew, 3, 4, 14-17, 23, 25, 32, 42, 49, 50, 53, 75, 79, 80, 84, 86, 88, 125, 130, 137, 140, 145

case study, 28, 35, 69-76, 79, 80, 86, 88, 110, 128, 132, 153.

company resource management, 4

competence, 3, 7, 8, 22, 31, 33, 35, 45, 52-56, 61, 61, 66, 109, 133, 138

crew resource management, 3, 6, 9, 10, 37, 61, 91, 167, 168

generations of CRM, 4-6, 68

critical incident, 48, 50

CRM. see crew resource management

culture, 9, 17-23, 26, 30, 41-45, 55, 56, 72, 110, 168, 169

curriculum, 62

debriefing, 69, 77, 81, 88, 95, 115, 117, 157

defences, 13, 27

error management, 4-7, 21, 27.

evaluation, 119, 123, 126-134, 136-138, 140-143, 162, 169

expertise, 5, 23, 35-38, 89, 130

FAA 3, 4, 6, 35, 63, 91, 107, 143, 166

grade scale, 148-153, 155, 163, 164,

human factors, 3, 5-7, 9-11, 91, 142

JAA. see joint aviation authorities

joint aviation authority, 6

knowledge, 5, 23, 25-28, 33, 35-38, 44, 54. 65, 68-71, 74, 87, 90, 115, 119, 136, 138, 147, 148, 158

lecture, 5, 69-72, 112

lesson, 23, 25, 28, 29, 32, 49, 62, 66, 67, 69-73, 76, 11, 109-113, 117, 127, 129, 132, 153, 162

lesson plan, 62, 66, 67, 107, 109, 112, 119

line operational safety audit, 24

LOFT, 69-72, 87, 88-90, 117, 132, 149, 150

managing change, 162

marking schemes, 46

non-technical skills, 45-46

objectives, 9, 51, 62-64, 66, 68-70, 76, 88, 106, 109-113, 118, 138

observer bias, 147, 151-153, 156,
    163, 164
organisations, 8, 11, 22, 23, 26, 27,
    30, 40, 43, 54, 56, 62, 73, 106,
    130, 141

practical exercise, 69, 80-82, 90
problem individuals, 115, 116

question technique,115-117
questionnaire, 42, 65, 69-72, 85-87,
    104, 114, 129, 131, 142

recalibration, 163, 164, 170
reliability, 47, 50, 144, 147, 148,
    151, 154, 165, 168, 169
remedial training, 164, 167
repertory grid, 48, 49
return on investment (ROI), 132,
    133, 136, 141
risk, 6, 11, 13-17, 21, 22, 25, 27-30,
    33, 37, 40, 41, 44, 45, 52, 54, 55,
    64, 67, 75, 83, 86, 89, 90, 110,
    117, 133, 137, 138, 139, 150, 152,
    154
role-play, 71
rule, 21, 34, 37, 39, 48, 61, 71, 73,
    77, 109, 119, 127, 151, 157

safety, 5-11, 13, 17-32, 37, 39, 41-
    43, 45, 53-55, 68,72, 82, 88, 89,

91,  106, 107, 118, 123, 125, 129,
    131, 135, 136, 138, 142, 143, 145,
    152-154, 159, 162, 165, 168, 169
safety culture, 17-21, 30, 41, 42,
    72, 110, 168, 169
safety management systems, 24, 28
skill, 4, 6-8, 22, 23, 31-34, 37-40,
    45-47, 50-54, 64, 66, 69-71, 73-
    75, 77, 80, 84, 85, 87-91, 109,
    115, 117-119, 123, 29, 126, 129,
    133, 137, 138, 140, 143-145, 148,
    150, 151, 153, 157-159, 162-165,
    167-171
sociological models, 7

standardisation, 26, 46, 123, 170
syllabus, 62, 63, 68, 80, 89, 107,
    168, 170

technical skills, 31-33, 45, 46, 52,
    144, 163, 165, 167
threat and error management, 21, 27

United Airlines, 3
University of Texas, 5, 46

validity, 47, 50, 69, 72, 78, 81, 88,
    146, 150, 165
video, 69-72, 74, 76, 113, 114, 132,
    134, 135, 145, 147, 163, 164